CITIES AND DISASTERS

CITIES AND DISASTERS

Edited by DAVIA C. DOWNEY

CRC Press
Taylor & Francis Group
Boca Raton London New York

CRC Press is an imprint of the
Taylor & Francis Group, an **informa** business

CRC Press
Taylor & Francis Group
6000 Broken Sound Parkway NW, Suite 300
Boca Raton, FL 33487-2742

Printed on acid-free paper
Version Date: 20150706

International Standard Book Number-13: 978-1-4822-4740-4 (Paperback)

Visit the Taylor & Francis Web site at
http://www.taylorandfrancis.com

and the CRC Press Web site at
http://www.crcpress.com

CONTENTS

SECTION III Resilience, Cooperation, and Citizen Attitudes

SECTION IV Planning and Path Dependence

PREFACE

DISASTERS: A CONCEPTUALIZATION

Disasters are "any event causing great harm, damage, or suffering" (Farazmand 2001, 2014). Birkland (2006) and Faulkner (2001) have defined disasters as crises (smaller-scale disasters with a relatively small scope of effects on a population), disasters (medium- to large-scale events of natural or human cause), and catastrophes (profound events that affect a broad range of land area and effectively render state and local governments unable to respond effectively). In the context of communities, natural disasters are accidental and can range from tornadoes, hurricanes, and earthquakes to heavy flooding, tsunamis, and droughts that affect the physical, social, and economic capabilities of a place.

Disasters can also be caused by human factors. Warfare, failures of technology, and human error, either accidental or purposeful in cause, can be just as devastating. Local, state, and national governments, nonprofit organizations, local welfare groups, religious organizations, the media, and other groups may be activated postdisaster to assist individuals and businesses with evacuating, stabilizing, and rebuilding. Thus, dealing with disasters often bridges many policy domains and requires both vertical and horizontal integration (Berke et al. 1993) and cooperation between various actors and agencies.

Fundamentally, responding to disasters takes time and oversight. The process of rebuilding, restoring order, and returning to what constitutes normalcy can take weeks, months, or even years, and communities are often left with permanent scars of these events. Disasters affect the physical environment, the political landscape, and the emotional well-being of the population. Planning and preparing for disasters are tasks that must be undertaken by actors big and small to reduce casualties and damage to infrastructure and to create the necessary partnerships with intergovernmental, nongovernmental, and other private actors and the public in order to make assets available that will need to be utilized in times of crises.

PUBLIC POLICY AND DISASTER PLANNING

Disasters are chaotic and can be difficult to manage due to the issues of scale, expertise, and physical vulnerability that are caused by these events. Emergency management, sometimes referred to as disaster management, is the process of creating plans whereby communities reduce susceptibilities to hazards and manage the process of recovery. Disaster management is a field of study that has generated much interest in recent years, as worldwide events of terrorism, large-scale natural disasters, and the effects of a growing world population have increased the number of incidents needed to be dealt with at the local, state, national, and international levels.

Generally, disaster planning consists of recognizing four phases: mitigation, preparedness, response, and recovery, with certain activities attributed to each. The first phase, mitigation, is concerned with government officials and other stakeholders deciding what to do when and if there is a risk to human safety and health. Mitigation is the stage where a hazard plan is developed to deal with different types of disasters. These plans identify, measure, and usually contain measures to reduce or lessen the likelihood of future repeated events. This might include restricting housing from being built in areas prone to flooding, requiring buildings to use certain materials in their construction to reduce the likelihood of damage during an earthquake, implementing new zoning laws to restrict building height, identifying floodplains, or requiring property owners to purchase hazard insurance in certain areas that are prone to natural disasters. After disasters, governmental organizations often update these mitigation plans to bolster response in future events (Sylves 2008).

The preparedness stage is when stakeholders develop a response plan and train first and secondary responders to save lives as well as reduce further damage. This stage also involves identifying needed critical resources, informing the local population of evacuation routes, developing service-sharing agreements with other agencies, creating mutual aid agreements, and readying the local population for emerging or expected threats (Haddow et al. 2011). The success of a preparedness plan is dependent upon the ability of stakeholders to identify potential hazards.

The response stage is where the mitigation plan and preparedness action steps are activated. The ability of local, state, federal, and nongovernmental groups to get to the affected area and follow the plan in a reasonable period of time is critical to reducing secondary damage, saving lives and property, and overcoming the event that has disrupted the community. Vertical coordination, which is the type of coordination that occurs between different levels of government, with the highest level (in the case of the United States, the federal government) taking precedence in directing the activities of lower-level government actors, is essential to ensuring that response is swift, particularly in instances where local-level actors are subsumed during a disaster. Horizontal coordination, which is the type of coordination between actors within the same level of government or community, can also predict the likelihood that the recovery stage is surefooted. Both horizontal and vertical coordination are necessary for disaster response to be swift, as noted by many scholars (Birkland 1998; Comfort 1993; Schneider 1995). The response stage is also the one most fraught with tension and often can be politically troublesome for government officials who fail to act quickly and decisively during a disastrous event (Waugh 2006). Hurricane Katrina in the United States is an example of how the response stage can be complicated by ineptness and inaction (Vogel 2009; Walsh 2010).

The recovery stage, the last stage of the emergency management cycle, is the most expensive, as it pertains to the actual rebuilding of assets, returning the local population to their homes (or in some cases, relocating them permanently), and returning to a level of normalcy. This stage can take months or years to rebuild infrastructure, and for many scholars, this stage is the hardest to quantify, as communities are often permanently changed by these events (Kettl 2006; Liu et al. 2006; Thomas 2010; Webb et al. 2002; White and Eisinger 2006). Once a place is deemed to be recovered, the process of updating hazard mitigation plans, bolstering vertical and horizontal partnerships, and working to alleviate the structural, political, or social rifts exposed during the disaster continues in earnest so that if a new event occurs, the community will be ready.

EMERGENCY MANAGEMENT: A FRAMEWORK FOR GOVERNMENTS

The United States

In the United States, public officials pay attention to emergency management particularly when highly visible events are handled poorly (Howitt and Leonard 2009). However, in the United States, as in most developed countries, local and state officials engage in emergency management and handle incidents well. Considering the range of natural disasters that have been tracked annually by the Federal Emergency Management Agency (FEMA) since 1953, the highest number recorded was 242 in 2011, with

111 of those being wildfire-related (FEMA 2014). National disaster declarations are rare; only when governors request a declaration to the president via the regional FEMA office are national assets such as funding, personnel, and other assistance given. Many disasters are managed either locally or by the state alone. This type of bottom-up response is most effective when horizontal and vertical policy coordination are utilized in an effective way (Peters and Pierre 2006). Local governments are close to disaster sites and usually have some emergency capacity to respond to disasters. Oftentimes local and state governments have mutual aid compacts that can be activated when the need arises.

The federal government has an emergency response plan apparatus that is activated during catastrophic events. Historically, the United States has been concerned mostly with ensuring that its public infrastructure is protected from attack. During the 1950s and 1960s, most public policy directed at emergency management was created to ensure that civil defense was impenetrable. President Nixon was the first to redefine the existing civil defense strategy to include preparedness for natural disasters. Thus, policy and planning shifted from structural mitigation (building dams and bridges) to nonstructural mitigation (the use of wetlands as natural buffers, etc.). With the passage of the Disaster Relief Act of 1974, the executive branch sought to streamline the hodgepodge of bureaucratic agencies tasked with acting during disasters (Schneider 1995).

When President Carter took office, he went one step further and initiated the birth of FEMA, a single entity responsible to the president for anticipating, preparing for, and responding to major civil emergencies (Howitt and Leonard 2009). During Carter's administration, FEMA did not fully consolidate all emergency management–related functions, so federal agencies continued to fight for jurisdiction over natural and man-made disasters until the passage of the Stafford Act of 1988, which amended the original 1974 legislation. The Stafford Act of 1988 enhanced presidential declaration authority. With the passage of the Stafford Act, the president is given more leeway in the types of disasters he or she can declare (Sylves 2008); the president can also direct federal disaster assistance to states and localities. The Stafford Act authorizes the president to issue a disaster declaration (if they are major events) and specify the type of assistance that will be given to the requesting state or area. These actions can include monetary compensation and grants, technical advice and assistance, and debris removal assistance. This update to the Stafford Act has allowed for much more political subjectivity in determining whether to declare a national disaster.

FEMA pre–September 11 was a relatively small agency; it did not directly provide emergency assistance but managed the operational response, relief, and recovery efforts of the federal government via presidential directives. FEMA's responsibilities were limited to assigning tasks to other governmental agencies and coordinating their work. FEMA also worked with local and state agencies, nonprofits, private contractors, and volunteer organizations to assist in relief efforts. After the September 11 terrorist attacks, however, the federal emergency preparedness organizational structure was overhauled, and FEMA became a part of the massive Department of Homeland Security (DHS).

With the establishment of the DHS in 2002, the United States revamped its emergency management and response plans, and the result was the National Response Plan (NRP), officially adopted in 2004. The NRP added two new categories of incidents beyond major disaster and emergency: catastrophic and National Special Security Events (NSSEs). Catastrophic events are "any natural or man-made incident, including terrorism, which results in extraordinary levels of mass casualities, damage or disruption severely affecting the population, infrastructure, environment, economy, and national morale and/or government functions" (Department of Homeland Security 2004, 42–44; King et al. 2009; Peters 2009). NSSEs are "high-profile, large-scale events that present high-probability targets."* The NRP, now renamed the National Response Framework (NRF), gives the president unencumbered authority to mobilize federal, state, and local resources during a disaster if it is concluded that an event could be construed as a potential act of terrorism. The creation of the DHS and FEMA's placement within the organizational structure of the DHS both have been seen as usurping FEMA's ability to act independently during other types of disasters (Ink 2008; Daniels et al. 2006).

International Emergency Management

Throughout the world there exist several organizations that deal with natural and man-made disasters: the International Emergency Management Society (IEMS), the International Federation of the Red Cross and Red Crescent, the World Bank, the United Nations, and the European Union. These organizations all work together during catastrophic events to assist local personnel in responding and recovering. The Red Cross and Red Crescent, IEMS, and other affiliated groups work to train first responders

* Presidential Threat Protection Act of 2000, public law 106–544.

and volunteers. The United Nations has a special office dedicated to emergency response within the Office for the Coordination of Human Affairs and will send teams of emergency responders at the request of affected countries. The World Bank has committed billions of dollars to various rebuilding projects worldwide in response to disasters. In most industrialized countries there exists a national organizational structure much like there is in the United States, with local government entities coordinating much of the initial response and the higher levels of government readying to assist when needed. Again, vertical and horizontal coordination among these actors is critical and can be tested by the same organizational and bureaucratic issues faced by emergency response teams in the United States (see Farazmand (2014) for a detailed description of each of these groups).

ORGANIZATION OF THE BOOK

Cities and Disasters seeks to provide a unique interdisciplinary view of emergency management policy and factors that affect the recovery of urban locations, populations, and the built environment after disasters strike urban and rural areas. The central theme that ties these contributions together is threefold: policy, urbanity, and the interplay by which disasters, whether they are man-made or natural, affect the process of communities returning to normalcy postdisaster. Some authors take a prescriptive approach, making policy recommendations to ensure that this area of study reflects the needs of communities, while others illustrate how disasters affect various aspects of society in cities. Contributors to this volume provide national and international perspectives on the methods various communities have used to approach the planning and recovery stages of disasters.

Section I: Marginalization and Recovery

In the first chapter, the effects of the marginalization of women during and after a disaster are discussed. The role of ensuring that women, in particular, are integral to economic and community recovery and the special attention that should be paid to their social vulnerability during the mitigation and preparedness stages of emergency and disaster management are important factors to consider when developing policy. In Chapter 2, the question of how nonprofit organizations in the United States provide a critical component to assisting communities in the response and recovery stages of disaster management is explored. In particular, the kind of nonprofits that serve vulnerable communities is discussed, and the situations these nonprofits face when thrust into the role of first responder postdisaster are highlighted. Chapter 3 examines how race and poverty have interacted with the economic recovery in urban and rural communities following Hurricane Katrina in the state of Louisiana.

Section II: Transportation Considerations

In the second section, issues of urban planning for civil defense are explored and analyzed as to how different types of transportation hubs need special attention, particularly during occurrences of warfare. Chapter 5 discusses the impact the port closure after Hurricane Katrina had on both the city of New Orleans and U.S. economic outputs, especially its effects on international trade. Both chapters in this section highlight the importance of understanding how disasters impact the ability of cities to recover critical assets that may be permanently damaged during catastrophic events.

Section III: Resilience, Cooperation, and Citizen Attitudes

Four contributions in this section highlight citizen attitudes and their interplay during the disaster recovery process. How citizens attribute blame and how blame affects governmental response are explored in Chapter 6 by looking at the aftermath of the *Deepwater Horizon* event. The aftermath of the Christchurch earthquakes in 2011 and how well the implementation of a local citizen engagement process impacted overall political participation in the postdisaster period are analyzed in Chapter 7. The importance of networks during the response phase of disasters is highlighted in Chapter 8. The authors' work identifies the interdependencies that emerge during the response and recovery phases and explores how gaps in collaboration have affected recovery in Joplin, Missouri. The final chapter in this section discusses the importance of understanding the effects of repeated displacement in areas prone to disasters and how planners should be attuned to the specific needs of vulnerability when assessing the efficacy of their emergency plans.

Section IV: Planning and Path Dependence

The final section discusses various issues of local land use and housing planning policies prior to and postdisaster. Chapter 10 utilizes the historical aspects of land use planning to illustrate how relegating blacks in

New Orleans to the Lower Ninth Ward from the very founding of the city resulted in catastrophic devastation during Hurricane Katrina. This chapter discusses the need for public officials to understand the intricacies of social vulnerabilities in the built environment and the special attention that these areas will need in the postrecovery period. Chapter 11 examines the importance of participatory planning in the postrecovery period in New Orleans and focuses on gentrification as a factor that has both positive and negative impacts on the sociological makeup of neighborhoods. Finally, Chapter 12 compares the relocation processes in Japan and the United States in the aftermath of two disasters (the Great East Japan Earthquake and Hurricane Katrina) and uncovers how national housing policy in the postdisaster context affects the socioeconomic makeup of cities after catastrophic events.

CONCLUDING THOUGHTS

This book seeks to add to the growing literature on emergency management and disaster planning and hopes to contribute some unique perspectives on disasters that have not yet been presented. Contributors to this volume explore various case studies on the economic, planning, and policy issues that have emerged in communities in the United States and the world before, during, and after disasters. There is much to be gained by emergency management scholars, students, first responders, and policy makers involved in the field of emergency management, and it is my hope that the conclusions drawn by the contributors of this book help inform decision makers. This area of policy continues to grow and expand specifically so that political actors can improve upon the ways in which communities are able to respond to unplanned events. As these policies are developed at the local, national, and international levels to mitigate the effects of future catastrophes, it is my hope that the analyses contained in this volume will help actors make decisions that strengthen communities for the future.

Davia Cox Downey
Editor

REFERENCES

Berke, P. R., J. Kartez, and D. Wenger. 1993. Recovery after disaster: Achieving sustainable development, mitigation and equity. *Disasters* 17(2):93.

Birkland, T. A. 1998. Focusing events, mobilization, and agenda setting. *Journal of Public Policy* 18:53.

Birkland, T. A. 2006. *Lessons of Disaster: Policy Change after Catastrophic Events*. American Governance and Public Policy Series. Washington, DC: Georgetown University Press.

Comfort, L. K. 1993. Disastrous management. *Public Administration Review* 53(2):179.

Daniels, R. J., D. F. Kettl, and H. Kunreuther. 2006. *On Risk and Disaster: Lessons from Hurricane Katrina*. Philadelphia: University of Pennsylvania Press.

Department of Homeland Security. 2004. *Final Draft: National Response Plan*.

Farazmand, A. 2001. *Handbook of Crisis and Emergency Management, Public Administration and Public Policy*. New York: Marcel Dekker.

Farazmand, A. 2014. *Crisis and Emergency Management: Theory and Practice*. 2nd ed. Boca Raton, FL: CRC Press.

Faulkner, B. 2001. Towards a framework for tourism disaster management. *Tourism Management* 22(2):135.

FEMA. 2014. About the agency. Washington, DC: Federal Emergency Management Agency, U.S. Department of Homeland Security. https://www.fema.gov/about-agency [last modified January 31, 2015].

Haddow, G. D., J. A. Bullock, and D. P. Coppola. 2011. *Introduction to Emergency Management*. Burlington, MA: Elsevier Science.

Howitt, A. M., and H. B. Leonard, eds. 2009. *Managing Crises: Responses to Large-Scale Emergencies*. Washington, DC: CQ Press.

Ink, D. 2008. Crisis management. http://transition.napawash.org.previewc40.carrierzone.com/papers/Ink_Crisis_Management_08_04_08.pdf.

Kettl, D. F. 2006. Is the worst yet to come? *Annals of the American Academy of Political and Social Science* 604:273.

King, P. T., M. Souder, D. Lungren, M. Rogers, D. Reichert, M. McCaul, and C. Dent. 2009. Keep FEMA within Homeland Security; splitting off agency courts disaster. *Washington Times*, January 14.

Liu, A., M. Fellowes, and M. Mabanta. 2006. *Katrina Index: Tracking Variables of Post-Katrina Recovery*. Washington, DC: Brookings Institute, Metropolitan Policy Program.

Peters, K. M. 2009. Moving FEMA out of Homeland Security could pose risks. *Government Executive*, February 17.

Peters, K. M., and J. Pierre. 2006. *Handbook of Public Policy*. Thousand Oaks, CA: Sage.

Schneider, S. K. 1995. *Flirting with Disaster: Public Management in Crisis Situations, Bureaucracies, Public Administration, and Public Policy*. Armonk, NY: M. E. Sharpe.

Sylves, R. 2008. *Disaster Policy and Politics*. Washington, DC: CQ Press.

Thomas, M. 2010. Hurricane Katrina, five years later. *New York Times Upfront*, May 10.
Vogel, S. 2009. GAO cites FEMA preparedness gaps: Agency making progress, report says. *Washington Post*, June 8.
Walsh, B. 2010. The Gulf disaster: Whose asses need kicking? *Time*, June 10.
Waugh Jr., W. L. 2006. The political costs of failure in the Katrina and Rita disasters. *Annals of the American Academy of Political and Social Science* 604(1):10. doi: 10.1177/0002716205284916.
Webb, G. R., K. J. Tierney, and J. M. Dahlhammer. 2002. Predicting long-term business recovery from disaster: A comparison of the Loma Prieta earthquake and Hurricane Andrew. *Environmental Hazards* 4:45.
White, A., and P. Eisinger, eds. 2006. *Urban Conversations: Cities at Risk: Catastrophe, Recovery and Renewal in New York and New Orleans.* New York: Milano, The New School for Management and Urban Policy.

EDITOR

Davia Cox Downey is the MPA coordinator and an assistant professor of public administration at Grand Valley State University, Allendale, Michigan. She teaches undergraduate and graduate courses in local politics, public policy, and public administration and has authored and coauthored journal articles, book reviews, and book chapters. Her most recent publication, "Civic Culture in Ottawa: The Endurance of Local Culture," coauthored with Laura A. Reese and Raymond Rosenfeld, can be found in *Comparative Civic Culture* (Ashgate Publishing, 2012). She has also written "Disaster Recovery in Black and White: A Comparison of New Orleans and Gulfport," published by the *American Review of Public Administration* in 2014.

CONTRIBUTORS

Michael O. Adams is the interim chair of the Department of Political Science, director of the NASPAA-accredited master of public administration program, and founding director of the online executive master of public administration program at Texas Southern University. He also serves as the interim director of the Barbara Jordan Research Institute. He is a tenured professor, teaching classes in public administration and political science, while continuously contributing to the academic body of knowledge through research. Dr. Adams has received numerous awards, including the Ford Foundation Doctoral Fellowship and the Committee on Institutional Cooperation Fellowship, the Barbara Jordan–Mickey Leland School of Public Affairs International Service Award, the Texas Southern University Research Scholar Award, the School of Public Affairs Research Scholar Award, and the Quiester Craig Teaching Award. He is an alumnus of Tougaloo College, Mississippi, where he earned his undergraduate degree. Dr. Adams earned his master of arts and PhD in political science from the Clark Atlanta University, Georgia.

Brittany N. Arceneaux grew up in the New Orleans area, where she dedicated her time to service and the community. After graduating from Cabrini High School, she moved to Austin, Texas, to attend St. Edward's University. Arceneaux soon returned to New Orleans to pursue a bachelor's degree in urban planning at the University of New Orleans (UNO). During her time at UNO, she worked with a group of professional and student volunteers as the program operations manager of WhoData.org. After graduating with a bachelor of science in urban studies and planning, Arceneaux began working at Pond & Company as a planner and geographic information systems (GIS) analyst. She maintains her relationship with WhoData.org as a volunteer assistant director. Her roles in project management include work in neighborhoods such as the Lower Ninth Ward and Hoffman Triangle. Along with filed research, Arceneaux works with neighborhood organizations to provide GIS training and methods of data collection. She has presented her research at conferences including the 2012 New Orleans Housing Community Development Conference and the 2012 National Reclaiming Vacant Properties Conference. She has dedicated years to community technical support and outreach. Arceneaux's work has impacted projects with numerous organizations throughout the New Orleans region.

Jane Carr is a doctoral candidate and instructor in public administration at Portland State University, Oregon. She teaches introductory civic leadership as well as courses on the roles of social institutions and individuals in civic engagement. Her research interests include theories of civic capacity and new governance, as well as the intersections of these frameworks with higher education. Carr earned a bachelor's degree in history and a master's in education.

Grace L. Chikoto is an assistant professor of nonprofit management at the University of Wisconsin–Milwaukee. She teaches courses in nonprofit management and public administration. Her research interests include implications of funding on nonprofit organizations, state–nonprofit relations, and nonprofit emergency management practices.

Antoinette S. Christophe is an assistant professor in the Executive Masters of Public Administration (eMPA) Program at Texas Southern University's (TSU) Barbara Jordan–Mickey Leland School of Public Affairs (BMLSPA). Dr. Christophe is a scholar, researcher, and educator in online distance learning as well as traditional classroom instruction. Her dissertation research, entitled "FEMA, EPA and Red Cross Response to Katrina," was published in the American Society for Public Administration's (ASPA) newsletter *SERENA* in January 2009. She is the lead author of the chapter entitled "Management of Hazardous Chemicals during Natural Disasters" in Dr. Ali Farazmand's book *Crisis and Emergency Management: Theory and Practice*. She is a coauthor of the article entitled "Graduate Education: Teaching and Advising Graduate Students Track Summaries," which appeared in *PS: Political Science and Politics* in 2013. Dr. Christophe has presented her research numerous times at various professional conferences over the years.

Dr. Christophe earned a doctorate in public policy; two master of science degrees, one in environmental chemistry and the other in analytical chemistry; and a bachelor of science degree in microbiology. She attended Michigan Tech University as an exchange student, where she received a certificate in sustainable development from the Sustainability Institute. She was an associate Western University fellow in 1995, a National Science Foundation IGERT fellow from 2005 to 2008, and is presently a National Science Foundation IGERT fellow alumnus.

Dr. Christophe was an active practitioner in the physical and social sciences for more than 25 years before becoming an academic. She was employed in the chemical industry for 15 years as a chemist and the Louisiana environmental regulatory industry for 6 years as an environmental scientist. Her work experience also includes duties as an assistant supervisor at Louisiana State University's Public Policy Research Lab for 2 years and training with the Louisiana Legislative Auditors as an intern for the Hurrican Katrina Road Home Program. Dr. Christophe uses her practical experience to enhance the learning experience for students in the eMPA and MPA programs at Texas Southern University.

Bridgette Cram is a PhD candidate in the Department of Public Administration at Florida International University. Her dissertation research will focus on cultural competency operationalization and assessment. In addition, her current research also focuses on social equity, gender studies, and higher education policy.

Ann-Margaret Esnard is a professor in the Andrew Young School of Policy Studies at Georgia State University (GSU). Her expertise encompasses urban planning, disaster planning, hazard and vulnerability assessment, and geographic information systems (GIS)/spatial analysis. Esnard has been involved in a number of related research initiatives, including National Science Foundation–funded projects on topics of population displacement from catastrophic disasters, and long-term recovery. She is the coauthor of the book, *Displaced by Disasters: Recovery and Resilience in a Globalizing World* (2014), published by Routledge Press. Esnard has served on a number of local, state, and national committees, including the Steering Committee for Evaluation of the National Flood Insurance Program, the Disasters Roundtable of the National Academy of Sciences, the National Research Council's Committee on Private–Public Sector Collaboration to Enhance Community Disaster Resilience, and the State of Florida Post-Disaster Redevelopment Planning Initiative.

Jean-Claude Garcia-Zamor is a professor of public administration at Florida International University. From 1999 to 2013 he conducted research and taught at Leipzig University in the former East Germany for fifteen consecutive summer semesters. In June 2007, the president (*rektor*) of that university designated him an honorary professor (*honorarprofessor*). Dr. Garcia-Zamor is the author of six books and editor of three volumes. He also coedited three other books. His twelfth book, entitled *Planning*

Strategies for Urban Development in Leipzig, Germany: Harmonizing Planning and Equity, was published in 2014 by Springer Publishing. Dr. Garcia-Zamor earned a PhD in public administration from New York University.

Mahshid Ghorbanian currently works as an assistant professor of urban design at the University of Art, Tehran, Iran. She earned her PhD degree in geography and urban planning from Tabriz University, Tabriz, Iran, in 2013. She is the author or coauthor of more than 18 scholarly articles and monographs, and author or editor of 5 books, including *City Center: From Theory to Urban Design Guidelines* (2014), *Urban Districts in Iran* (2006), and *Urban Climatology* (2005). Dr. Ghorbanian's research interests include urban morphology and urban design techniques.

Jillian Girard is a doctoral candidate at Portland State University, Oregon, in the Public Affairs and Policy Program. She has worked as an evaluator and research assistant for projects with local jurisdictions in and around Portland for the past three years. Additionally, she is the program coordinator for two graduate-level fellowship programs, the Oregon Fellows and the Hatfield Resident Fellows, collectively known as the Next Generation Initiative. Girard is also an active member of Portland City Club and the Oregon Program Evaluators Network. Her research interests include natural resources, civic engagement, and genetically modified food.

Stephanie Hawke is a doctoral candidate in public affairs and policy in the Mark O. Hatfield School of Government at Portland State University, Oregon. Her research interests include issues of race and representation in urban environments, as well as the political ramifications of disaster. In addition to pursuing her own research, Hawke is a consultant researcher with PSU's Center for Public Service and a political science instructor.

Michael Howell-Moroney is a professor and chair of the Department of Public and Nonprofit Administration at the University of Memphis, Tennessee. His research has been published in numerous scholarly journals in the fields of public administration and urban affairs. In addition, he has had many extramurally funded research projects, totaling over $3 million, including two grants from the National Science Foundation. He has a record of engaged scholarship, at both the local and national levels. He has served as an advisor on numerous boards, and has written or contributed to many reports and local community–based research projects. He recently coauthored a report for the International City–County

Management Association (ICMA) on the state of intergovernmental relations in the United States.

David E. A. Johnson is an associate professor in the Political Science Department of Missouri State University. He has authored numerous articles in emergency management based on his research focusing on vulnerability and the use of technology. He is a certified emergency manager, and prior to entering academia, he served more than 30 years in various public safety positions at the local, state, and federal levels. Today, in addition to his academic duties, he continues to serve with a regional incident management team and a federal incident response coordination team.

Kent R. Kerley is a professor and chair in the Department of Criminology and Criminal Justice at the University of Texas at Arlington. His primary research interests include corrections, religiosity, and intimate partner violence. He is the author of *Religious Faith in Correctional Contexts* (First Forum Press/Lynne Rienner Publishers, 2014) and the edited collection *Policing and Program Evaluation* (Prentice Hall, 2005). His research has appeared in many top journals, such as the *Journal for the Scientific Study of Religion, Justice Quarterly, Social Forces,* and *Social Problems.* He is principal investigator for two National Science Foundation undergraduate research training grants used to create a Research Experiences for Undergraduates (REU) program at UAB called Using the Social Sciences, Natural Sciences, and Mathematics to Study Crime (NSF Awards 1261322 and 1004953). He has also received research funding from Google and the Religious Research Association.

Tamiyo Kondo is an associate professor at the Graduate School of Engineering, Kobe University. She teaches housing policy and postdisaster recovery planning. Her areas of expertise include community-based planning and housing recovery after disaster. She has studied housing recovery after Hurricane Katrina (2005) in the United States and the Great East Japan Earthquake (2011), on which she coauthored the following articles: "Housing Recovery by Type of Resident Involvement—Providing Housing vs. Mobilizing Residents" (*Proceedings of the International Society of Habitat Engineering and Design,* October 2012), and "Housing Recovery for Sustainable Disaster Recovery: Through Case Study of Hurricane Katrina (2005) and Great East Japan Earthquake (2011)" (extended abstract of the 3rd International Conference on Urban Disaster Reduction, October 2014).

Grace Elizabeth Major is an undergraduate in the Urban Studies Program at New College of Florida, concentrating on urban geography and public participation geographic information systems (GIS). She attended classes at the University of New Orleans's Planning and Urban Studies Department in spring 2014, where she got involved with Dr. Michelle Thompson and the WhoData.org project. She completed her undergraduate thesis on authenticity in the Tremé neighborhood in New Orleans in the spring of 2015. She plans to pursue a graduate degree in geographic information science in the fall of 2016.

Alka Sapat is an associate professor of public administration at Florida Atlantic University. Her expertise includes disaster and crisis management, postdisaster housing policy, environmental policy, and methodology. She was a research fellow with the National Science Foundation–funded Next Generation of Hazards Researchers program. Dr. Sapat has been involved in many NSF-funded projects on topics of building code regulation, disaster-induced population displacement, and the role of nongovernmental organizations in disaster recovery and resilience. Her work has been published in *Natural Hazards Review, Public Administration Review, International Journal of Mass Emergencies and Disasters,* and other scholarly publications. Her teaching interests include disaster management and homeland security, disaster planning and public policy, research methods, and statistical analysis. Dr. Sapat serves on the Florida State Disaster Housing Task Force and the Governor's Hurricane Conference Committee, along with serving on local committees on postdisaster housing initiatives. She is the coauthor of a recently published book, *Displaced by Disaster: Resilience and Recovery in a Globalizing World* (Routledge Press, 2014), and editor of a forthcoming book, *Post-Disaster Housing* (to be published by CRC Press in 2016).

Amir Shakibamanesh currently works as an assistant professor of urban design and vice chancellor for research at the University of Art, Tehran, Iran. He earned his PhD degree in urban design from Iran University of Science and Technology, Tehran, Iran, in 2013. He is the author or coauthor of more than 25 scholarly articles and monographs, and author or editor of 7 books, including *City Center: From Theory to Urban Design Guidelines* (2014), *Urban Design with Passive Defense Consideration* (2012), and *Urban Districts in Iran* (2006). Dr. Shakibamanesh's research interests include passive defense and virtual reality.

Michael D. Siciliano is an assistant professor in the Department of Public Administration at the University of Illinois at Chicago. He earned a master's degree in policy analysis from Carnegie Mellon University and a PhD in public policy and public administration from the University of Pittsburgh. Dr. Siciliano's current research connects network theory and complexity theory to ongoing problems in public policy and public administration. He is particularly interested in the formation of social and professional networks within public organizations and the implications of these networks on individual and organizational behavior and performance. His work also examines the interorganizational networks that emerge in response to disaster events and the factors influencing nonprofit collaboration in local service provision.

Michelle M. Thompson is an associate professor at the University of New Orleans (UNO) in the Department of Planning and Urban Studies (PLUS). She teaches courses in geographic information systems, community development finance, urban public finance, housing, urban studies, and land use planning. She earned her master's in regional planning (1984) and PhD (2001) from the Cornell University Department of City and Regional Planning, with a focus on community development and spatial analysis using geographic information systems (GIS). Dr. Thompson earned her bachelor of arts in policy studies from the Maxwell School of Citizenship and Public Affairs at Syracuse University in 1982.

Her research focuses on the application of public participation geographic information systems (PPGIS) in community development and reinvestment. She is the project manager of the web-based community mapping service WhoData.org, which combines parcel-level neighborhood condition information with public data to monitor socioeconomic and demographic changes.

Dr. Thompson is also the principal of Thompson Real Estate Consultants LLC, a real estate research and education firm. She has worked in both public and private companies related to the financing of residential and commercial real estate. She has a long-term interest in working with community development organizations to provide technical support, market research, and evaluation services.

Nada Toueir is currently a PhD candidate in the Faculty of Environmental Design at the University of Montreal and is planning on finishing her degree in the fall of 2015. Her research focuses on the resilience of urban communities in postdisaster situations, and specifically on the Lower

Ninth Ward in post-Katrina New Orleans. She emphasizes the cultural landscape and intangible elements such as place attachment, social networks, and memory in understanding how they contribute to people's resilience. Toueir's current research targets the most vulnerable and most devastated areas and populations, with an emphasis on social injustice. Her research interests were inspired by personal experience in her home country of Lebanon during the war of 2006. Later, through her work with Cornell University on developing a rebuilding plan for the Ninth Ward in New Orleans, her doctoral project took form.

Clayton Wukich is an assistant professor in the Department of Political Science at Sam Houston State University. He teaches courses on public management, organizational theory, public policy, and state and local government. His research interests include intergovernmental relations and collaborative governance, particularly in the policy domain of disaster and emergency management. His work has been published in *Public Administration Quarterly; Risk, Hazards and Crisis in Public Policy;* and the *Journal of Public Affairs Education;* among others.

Section I

Marginalization and Recovery

1

Enhancing Community and Economic Development Postdisaster through the Increased Resilience of Women

Bridgette Cram and Jean-Claude Garcia-Zamor

Contents

INTRODUCTION

Postdisaster development is of critical concern to disaster researchers and policy makers. This chapter examines the important role that resilience, especially women's resilience, plays in fostering postdisaster community and economic development. The chapter reviews relevant literature from several perspectives and then presents a discussion of how resiliency policy is being implemented across the globe, in addition to specific policy recommendations that can be applied to increase resiliency and foster community and economic development. The focus on women's resilience is necessary, due to the fact that if given the proper access and support, they enhance the recovery and development process. For the purpose of this chapter, several types of resilience will be discussed. Table 1.1 defines the three types of resilience that are imperative to this chapter: disaster resilience, community resilience, and economic resilience.

It is important to note that resiliency is the result of mitigation and preparedness efforts, as well as the combination of response and recovery plans. Each of the three definitions hints at how the combination of efforts in terms of mitigation and preparedness affects a community's ability to "bounce back" to its "new normal."

The chapter is significant because the topics discussed can be applied in an international context. No country, regardless of its gross domestic product (GDP) or level of power, is completely immune to the effects of a disaster. The concept of postdisaster resiliency is something that policy makers must focus on to not only find ways for communities to recover

Table 1.1 Operational Definitions of Resilience

Disaster resilience	"The ability of a social system to respond and recover from disasters and includes those inherent conditions that allow the system to absorb impacts and cope with an event, as well as post-event, adaptive processes that facilitate the ability of the social system to re-organize, change, and learn in response to a threat" (Cutter et al., 2008, p. 599).
Community resilience	"The capability to anticipate risk, limit impact, and bounce back rapidly through survival, adaptability, evolution, and growth in the face of turbulent change" (Plodinec, 2009, p. 10).
Economic resilience	"[Dynamic] resilience: the speed at which an entity or system recovers from a severe shock to achieve a desired state. "Static economic resilience: the ability of an entity or system to maintain function (e.g., continue producing) when shocked. "Inherent resilience: the ability to deal with crises. "Adaptive resilience: the ability (of an entity or system) in crisis situations to maintain function on the basis of ingenuity or extra effort" (Rose, 2007).

more quickly, but also set the stage for encouraging development postdisaster that fosters and enhances community and economic resiliency, leaving a community better prepared and more resilient than it was prior to the storm.

The central thesis of the chapter is that a higher level of community resilience is related to improved community and economic development postdisaster. More specifically, however, we will examine how the increased resilience of women as a component of community resilience will foster these postdisaster development goals. It is important to note that when the cited authors refer to normative gender roles, they are referring to how these roles have been defined in terms of "shared beliefs about desirable qualities for men and women in their society" (Berger and Krahé, 2013, p. 516). To achieve this, we first present a review of the literature that pertains to the effect of resilience of community and economic development, followed by a discussion of women and what makes them more or less resilient in the disaster context. Next, we discuss how the evidence found in the literature is demonstrated through several case studies, which show how the increased (or decreased) resiliency of women can have an impact on the recovery process. We conclude with the policy implications that can be derived from our analysis, as well as recommendations made by several scholars and international agencies.

LITERATURE REVIEW

The concept of resiliency is integral to postdisaster community and economic development. The literature reviewed in this section will first explore the role of resilience in terms of community and economic development and will then examine the effect of women's resiliency in this same context. Articles focusing on both domestic and international disasters are analyzed. The literature review concludes with a synthesis of several case studies that explore the role of women in disasters.

Resilience and Community and Economic Development

The key to resilience after a disaster is highly dependent upon a community's levels of vulnerability and resilience predisaster. Several authors have noted the effects that social vulnerability has on a community's ability to be resilient. However, community leaders may have a difficult time understanding where their community is most vulnerable. Cutter et al. (2003) noted the difficulty in quantifying social vulnerabilities; they also made the observation that "social losses are normally absent in after-disaster cost/loss estimation reports" (p. 243) due to the frequent absence of their measurement. This presents a problem in terms of being able to understand how vulnerable a community is predisaster and what can be done to mitigate these vulnerabilities to enhance postdisaster resilience.

Understanding the existing vulnerability of a community is an important first step in fostering communities that are resilient in terms of community and economic development postdisaster. Although a community may be vulnerable, if taken advantage of, the postdisaster time period serves as an excellent starting point to put policies into place that enhance community resilience. As Pyles (2007) states, "a core and often neglected element of disaster recovery has been the rebuilding and community development phase" (p. 321). It is critical to examine how a community emerges from a disaster more resilient than it was before, leaving it more prepared and less vulnerable for the next event that may occur.

Factors That Compose Resilience

Resilience can be seen as a summation of proactive and reactive measures and adaptive strategies. According to an analysis completed by Berke et al. (1993), the redevelopment process is rarely predictable, requiring stakeholders to come up with adaptive strategies addressing whatever needs their community may have. Based on this need for adaptation, it is surprising that adaptive decision making is not one of the main focuses for emergency managers. Because of the flexibility needed after a disaster, it is important to "develop institutional arrangements for disaster recovery planning that foster rather than constrain learning" (p. 97).

The conceptual framework presented by Berke et al. (1993) demonstrates the importance of horizontal integration among "a community's various social units and subsytems" (p. 100). This horizontal integration is critical to the resilience of a specific community; communities with high integration demonstrate strong relationships among organizations and stakeholders, leading to a mutual "interest in public policy decisions" and the ability to communicate effectively and make decisions that will benefit the entire community. Those communities with low horizontal integration lack a tight social network, leading to weak partnerships among stakeholders and a fragmented approach to solving problems. The concept of vertical integration is also imperative, as this represents the "structural and functional relation of [a community's] various social units and subunits to extra-community systems" (Berke et al., 1993, p. 101). Thus, it is critical for communities that wish to be resilient to ensure that they have a high level of vertical and horizontal integration present before a disaster occurs. Recommendations for increasing this integration are presented later in the chapter.

Factors That Influence Community Resilience

When examining the concept of resilience, it is imperative to understand that there are several factors that influence a community's ability to thrive after a disaster. According to Bruneau et al. (2003), resilience is determined by mitigation efforts made by "social units (e.g., organizations, communities)" and institutions and organizations within the built environment, such as "water and power lifelines, acute-care hospitals, and organizations that have the responsibility for emergency management at the local community level" (p. 735). In addition to institutional and organizational impacts on resilience, the capability of individuals and groups of individuals plays an important role in postdisaster recovery and development, as "a supportive social context in a community, prior to an adverse event, has emerged as a key component of resilience and provides a bridge between individual resilience theory and an exploration of a community-level theory" (Plough et al., 2013). Thus, it is clear that in order to increase resilience, a multipronged approach is necessary.

Women and Disaster Resilience

Although research demonstrates that women face several vulnerabilities in the face of disaster (Enarson, 2012; Enarson et al., 2007), it is imperative to understand how they contribute to resilience and subsequent development efforts. A gendered lens is relevant to the study of disasters and emergency management due to the fact that "gender strongly influences the behaviors and experiences of men and women at all phases of the hazards cycle" (Tierney, 2006, p. 120). A critical factor to consider in the ability of women to be resilient and have a positive effect on postdisaster recovery and development is the presence of social capital. Social capital is defined as bonds, bridges, and linkages:

- Bonds: links to people based on a sense of common identity ("people like us")—such as family, close friends and people who share our culture or ethnicity
- Bridges: links that stretch beyond a shared sense of identity; for example distant friends, colleagues and associates
- Linkages: links to people or groups further up or lower down the social ladder (What Is Social, 2007)

This concept is vital when it comes to analyzing how disaster not only differentially affects each gender, but also may exacerbate already existing inequalities. Furthermore, Putnam (1993) defines social capital as "networks, norms, and trust that facilitate coordination and cooperation for mutual benefit. Social capital enhances the benefits of investment in physical and human capital" (p. 35). In disaster contexts, it is critical to identify how these elements can foster resilience. This section looks at examples from the disaster literature that have specifically explored women and social capital.

Social capital, both positive and negative, affects the resiliency of women (Ganapati, 2012). These effects have been studied little in terms of gender; however, it is noted that the presence of social capital predisaster can play an important role in disaster preparedness and response as well as "contribut[ing] to successful and speedy household and community recovery" (Ganapati, 2012, p. 412). Based on this observation, it seems clear that having a solid foundation of women's civic and social networks predisaster would enhance resilience after a disaster. Thus, it is critical to examine how these groups are formed and what considerations must be made to avoid the negative effects of social capital that will be introduced later in the chapter.

In terms of the creation of social capital, David (2010) presents an excellent review of how women came together, oftentimes through grassroots efforts, to assist in all areas of the disaster cycle. His review highlights the importance of previously formed social and friendship networks, as they provide women with the ability to have a set location and access to resources that foster the involvement of women throughout the disaster cycle. Further examples are drawn from Pyles and Lewis's (2007) work on the mobilization of several women's groups after Hurricane Katrina. Their research indicated that the grassroots or informal organizations that were created were critical; while formal organizational networks may ignore the important assets that indigenous volunteers bring, the inclusion of grassroots and information organizations, which are often led by or composed of women, can result in increased citizen participation, particularly by those in isolated or marginalized communities (p. 389).

Although social capital is important for all members of a community, it can be particularly important to develop this among women. Increasing women's social capital is critical, as generally their roles predisaster, as well as the roles they acquire during a disaster, can be enhanced by these networks. According to Tierney (2006), social capital is an important factor in terms of disaster recovery; she states that it serves as a vital resource and that "well-off disaster victims typically have skills that the poor lack, such as knowing how to access resources and navigate bureaucratic requirements successfully" (p. 115).

A Framework for Resilience

It is important to consider how increasing the resilience of women can be integrated into the factors that both compose and influence resilience in general. Tobin (1999) introduced a thought-provoking framework for analysis applied to the creation of sustainable and resilient communities. He combined three models previously advanced and described how the interaction of these models led toward a goal of sustainable development, which decreased risk and increased resilience. The three models—mitigation, recovery, and structural–cognitive—demonstrate the intertwined nature that planning for a disaster requires. Although the mitigation and recovery models are quite straightforward, it is the integration of the structural–cognitive model that appears to be an important link, but one that many leaders or politicians may not understand. This component combines any structural changes that need to happen within society and changes that are more situational in nature, as "physical, social, cultural, and economic factors may all constrain (or promote) action" (p. 15). Furthermore, Tobin recognized the importance that gender played in this context, as women many times faced "different consequences" than their male counterparts, thus necessitating their participation in the disaster planning process. It is clear that communities must pay attention to all stakeholders and address the underlying social as well as structural issues should they wish to be resilient in the face of a storm.

The review of the literature and application of the resilience framework provides a solid foundation for the remainder of the chapter. These concepts and themes will be expanded upon using multiple perspectives. The following case studies and policy recommendations will demonstrate how incorporating women and enhancing their resiliency can make an important impact on their overall community and economic development resilience postdisaster.

DISCUSSION AND CASE STUDIES

The review of the literature demonstrates why fostering resilient cities is of the utmost importance for postdisaster development. Furthermore, the United Nations has identified disaster resilience as a key component for its post-2015 development agenda. Essentially, policy makers and emergency managers must be cognizant of the role that well-prepared citizens, institutions, and organizations play in enhancing postdisaster community and economic development. We present four case studies that demonstrate how women demonstrated resiliency in a postdisaster context. Each case study will present a brief example from the related literature, and we then provide a synthesis of lessons learned from each.

Case Study 1: Hurricane Mitch, Nicaragua

The first case study is related to the effect of Hurricane Mitch (1998) on Nicaragua. Cupples (2007) identified several communities that were affected by Hurricane Mitch and discussed the dimensions of gender within each one. What is interesting about Cupples's research is that her goal was to identify the importance of "imparting a more gender sensitive framework that relief workers and agencies can deploy following a disaster" (p. 157). Her analysis of these communities in Nicaragua highlighted the lack of "gender-sensitive" relief that was provided. She linked the positive or negative experiences of the women she interviewed to their own personal experiences, to avoid generalizing that every woman experiences a disaster in a particular way.

The first example Cupples gave is the city of El Hatillo. Cupples described El Hatillo as a community that was relatively successful in the recovery process, something that she attributed to their predisaster levels of community cohesion. In terms of gender, she found that "the women in particular were organized into women's groups and participated in multiple forms of disaster work.... Indeed, the women of El Hatillo saw themselves as strategically reproducing normative gender relations in the interests of more long term emancipation" (Cupples, 2007, p. 160).

Although the women maintained these normative roles, they were still optimistic about the future, due to their focus on leveraging the development aid they received to work toward these higher-level goals of gender equity. The remaining examples that Cupples gave are not as positive, as the level of predisaster planning and community cohesion were not as strong as they were in El Hatillo. For example, in El Mirador, she found that the disaster aid disbursed led to "aid dependency and reinforcement of existing gender inequalities" (p. 160).

Case Study 2: Golcuk Earthquake, Turkey

The next case is the Golcuk earthquake that struck Turkey in 1999. Ganapati (2012, 2013) produced an extensive amount of research on the disaster, and one of her main focuses was on the gendered aspect of social capital during the disaster recovery process. Her research demonstrated both the positive and negative effects that social capital had on women.

Social capital proved to be extremely important for women in the post-earthquake context. According to Ganapati (2012), social capital in terms of the formation of women's networks helped women to better express themselves, "fight for their rights," "celebrate their identities as women," and "express their dissatisfaction with the gender bias in Turkish society in very subtle ways" (p. 423). Furthermore, these networks allowed women to "avoid the stigma of charity." Because of the gender roles present in Turkey, it was difficult for women to find work in order to support their families; they therefore needed to depend on government assistance to ensure that their family's needs were met. Ganapati found that the networks that emerged after the earthquake were extremely important to help women avoid this stigma and get the help that they needed. She found that the informal aid networks, which were especially important, were proactive in finding disaster survivors that needed aid and helped in whatever way they could to get this aid to them. Women who may have been afraid to seek assistance were able to get help through "word of mouth" and thus were able to "keep their pride and dignity" thanks to these aid networks (p. 424).

Although social capital had positive effects, Ganapati (2013) argued in a later article that it also had some negative effects, most notably that it portrayed women as "teary eyed" and put them in conflict with government officials. Her article is one of the first to tackle the downsides of social capital for women in the postdisaster context, and it demonstrated that due to existing gendered relations, women found it hard to enter into many of the more male-related networks that emerged after the earthquake, such as search and rescue teams.

Furthermore, Ganapati noted that after the earthquake, civic networks reinforced the portrayal of women as teary-eyed victims, using this representation of women to gain higher ratings in hopes of getting more support and "mobilizing their agenda" (p. 86). In terms of conflict with state authorities, Ganapati found that because Turkish women did not often interact with the government in negative ways, the interactions they had in terms of protests had a negative effect on them, subjecting them to extreme pressure.

Case Study 3: Grand Forks, North Dakota

This case study looks at the Grand Forks floods of 1997 in North Dakota. The research completed by Fothergill (1999) demonstrated that the floods could be examined from a gendered perspective in terms of the roles that women played postdisaster and their experience with receiving aid. In terms of women's roles during and after the floods, she found that women played an extremely important role in fostering the resilience of the community postflood. They engaged the community in several ways, including the "masculine" work of sandbagging, although their roles were generally restricted to gender norms, that is, preparing food for the sandbaggers, providing child and elder care for relatives and nonrelatives, and remaining responsible for their roles prior to the disaster (both family and career). Fothergill mentioned this tendency to place women into stereotypically female jobs that they would usually do in their own home to show that although they were fulfilling normative roles, since they were doing them in the public sphere, it was being recognized as important work and critical to the success of the community.

In addition to this community role, women also played the family and the work roles. Fothergill recognized that a woman's family role was the most "greedy"; however, women were eager to stay engaged in their other roles. They were valued and seen as important within their community, and "as a result, the self became more valued as well" (p. 140). Although women were generally restricted to traditional roles, Fothergill's research demonstrated that their role accumulation (and at times role conflict) was actually seen as positive, as it led to the women "see[ing] themselves differently as more competent, confident, and capable and worthy of multiple roles and responsibilities, especially in the public sphere" (p. 140).

Case Study 4: Northern Colorado Floods

The last case study examined the floods that occurred throughout northern Colorado in 2013. Based on several interviews with local community leaders and local officials, Cram (2014) examined the resiliency of women after these floods. In tight-knit, small, close communities, the level of resiliency was expected to be much higher. This is attributed to a high sense of community cohesion. In terms of gender, specifically, many attributed this community cohesion and the ability to organize and focus to the women of the community:

> What I found was that during the flood and after, it seemed that women were the ones who were comforting people, feeding and clothing people, making sure that people were OK. It was almost as if we were the moms to these individuals, people were relying on us: children, husbands, coworkers and friends. Women were really solid, they got it, they understood that we needed to go into crisis mode . . . women are so good at that because in society we are the caretakers, we are the ones people come to when they have problems and these women in [redacted] just gathered together and did it. I was extremely proud, the women came together and just said we'll worry about ourselves later and we are going to help people and help them get through this. I was very proud to see all of these women step up and make it happen (Cram, 2014, p. 6).

Although the gender roles here are also relatively traditional and stereotypical of female work, it is clear that women played an important role in the resilience of their communities. Additional interviews indicated that the population naturally self-segregated at the community level; men began the immediate process of cleaning and rebuilding, while women performed their roles as wives and mothers, but also as the ones responsible, more often than not, for navigating the bureaucracy of the aid process. In terms of this process, many interviewees acknowledged that women naturally formed groups to help one another through the process, spreading information via word of mouth and doing their best to help their neighbors in whatever ways they could.

Synthesis

Each of these case studies provides important insights into the complex role that gender plays throughout the disaster cycle. This synthesis will discuss these lessons learned, which will guide the recommendations presented at the end of this chapter. While interpreting the results of the case studies, it is important to take the concept of intersectionality into consideration. This concept refers to the intersection of gender with other identities (e.g., race, class) and how these intersections "contribute to unique experiences of oppression and privilege" (Symington, 2004, p. 1). Literature demonstrates that the social construction of vulnerability is tied to these intersecting identities; for women in particular, they face a higher likelihood of living in poverty, leading single-headed households, living in the economic margin, and being members of already marginalized racial/ethnic groups leading single-headed households and facing other social disadvantages (Donner and Rodríguez, 2008; Laska et al., 2008; Morrow, 1999). Thus, even for cases within the same country, it is not surprising that women with different intersecting identities will react to disasters in different ways.

For the case of Hurricane Mitch in Nicaragua, it is clear that the predisaster cohesion of the community in El Hatillo was critical to its recovery success. This is something that we have seen throughout the literature, and it demonstrates that predisaster capacity building is critical for a community if it wishes to be resilient after a disaster. Another important finding from Cupples's (2007) research is that aid had a negative effect on a community when not disbursed properly. In her example, we see how this led to dependency and an increase in gender inequality; it is imperative to think about how aid agencies can understand these negative effects and counter them by creating distribution schemes that enhance community cohesion to help the community become even stronger than it was before the disaster.

In the Golcuk earthquake case study, we were introduced to the role that social capital played in women's postdisaster experience. Both the negative and positive sides of social capital were presented, demonstrating that the emergence of civic networks either serves to enhance the role of women in society or has a foundation in the society's already existing gendered norms and exacerbates this inequality. Furthermore, a close eye needs to be kept on the media so that don't get away with the exploitation of women as victims for ratings and support.

In the example of the North Fork floods, we see that women were required to take on several roles in the postdisaster context. However, it is important to note that even though these women were still performing normative roles, they were also being held responsible for both their work and family roles. In some ways, engaging women in the public sphere postdisaster is seen as liberating for women, even if they are completing the same tasks they would have done inside their own homes. Based on the evidence that Fothergill collected, this seemed to be the case for the floods; however, if this increased recognition and enhanced sense of self served to benefit the women in the long run and allowed them continued access to the public sphere, it was not reported.

In the case of Colorado, we again see women taking on more normative gender roles. However, based on the interviews conducted, many of these roles were self-selected, and there was no evidence of women being excluded from any role they sought to take (i.e., search and rescue, cleaning, rebuilding). The reason for this seems to be connected to the extremely high cohesion of many of these communities; their predisaster level of community development was high, so when the floods happened, they were ready to join together and do what they needed to succeed.

Each of the case studies provides examples of women remaining in normative roles during the disaster recovery process. In the case of El Hatillo, we see that women attempted to use the disaster to enhance their equity through the opportunities given to them by the aid they received. From the Golcuk example, it is important to note that even though women were able to participate, it was imperative to ensure that they were allowed full participation regardless of gender. For the North Fork floods, this normative work seemed to be positive for both the community and the women, as it was recognized as being crucial to the recovery process. In addition, the North Fork floods women were more easily accepted into the "masculine" job of sandbagging because it was viewed as a critical part of the duty as a community member. In Colorado, although these roles were present, we see that they did not pose a significant problem for the communities, as the goal was to come together and get things done, regardless of whom was fulfilling which role.

Overall, these case studies present an important look into how gender affects an individual's role in the postdisaster context. It is clear that women tend to be pushed into more normative roles; however, the four case studies demonstrate the differing effects that these have on women. What is important to note is how these roles are distributed and viewed within society. If these roles are based on the current gender structure and women are excluded from work that is not considered feminine, then there is an important societal issue that must be addressed. However, if women, as in the example of Colorado, self-select into these roles, but are given equal opportunity to engage in what would be considered more masculine roles, then there is not a gender problem—as the roles are equally recognized as being important to the full recovery of the city.

The following "Recommendations" section takes the lessons learned from these case studies into account to develop practical advice that can be implemented in both developed and developing countries.

RECOMMENDATIONS

One of the most important ways that the resilience of women can be enhanced is through their inclusion in committees and the political sphere in terms of decision making. As demonstrated, women play a critical role in all stages of the disaster cycle, oftentimes accumulating additional roles that place additional stress on them. According to Laska et al. (2008), the world of emergency management is extremely masculine, often imitating the patriarchal top-down system that exists within family units; however, "many of the vulnerability and recovery needs of families, neighborhoods, and communities would be well served by engaging women as informants, responders, and emergency managers" (p. 19).

How can these recommendations be implemented at a practical level? On an international level, the importance of the gender perspective within the context of disasters is beginning to emerge. The Hyogo Framework for Action 2010–2015 (HFA) has an overarching goal of "the substantial reduction of disaster losses, in lives and in the social, economic and environmental assets of communities and countries"; to reach this goal, HFA recognizes a secondary crosscutting goal of gender perspective and cultural diversity awareness. However, the United Nations Office for Disaster Risk Reduction, which oversees the HFA, recognized in its midterm report that "inclusion of a gender perspective and effective community participation are the areas where the least progress seems to have been made" (p. 44). Therefore, it is clear that even though gender equity and community participation are recognized as critical factors to enhance resilience, making these changes and empowering women and communities has been a difficult task to achieve.

In formulating practical recommendations, it is important to recognize that even slight changes that empower women, improve community cohesion, or enhance predisaster community development can have a powerful effect on an individual's or community's ability to bounce back from a disaster. The following recommendations build upon the literature reviewed and the case studies presented. Specific policies related to both enhancing the resilience of women and improving community cohesion and development predisaster will be included. Furthermore, a discussion of postdisaster policies and procedures that help enhance the recovery process will be discussed.

Enhancing the Resilience of Women

The key to enhancing the resilience of women in the postdisaster context is closely tied to the preexisting gender structure in a community. This section will examine recommendations related to social capital and gender equity. Each of these concepts is critical to the enhanced resilience of communities as a whole.

Social Capital

We must address the issue of social capital as it relates to enhancing the resilience of women specifically. According to Ganapati (2012), "policy makers need to pay more attention to women's social capital and design more finely grained policies by taking gender into account in the context of disaster" (p. 424). Fukuyama's (2002) work on social capital is especially relevant within the context of development, as "the concept of social capital puts both policies and institutions in their proper cultural contexts" (p. 29); thus, the means to develop social capital will also vary depending on context. Fukuyama notes that generating social capital is most difficult in a society where there is a fundamental lack of social trust. In these instances, it is necessary to first focus on the political institutions and how they, and the law, can be strengthened. This needs to occur in addition to, or in conjunction with, promoting cooperation "between groups that typically have had little to do with one another" (p. 32). Most importantly, however, Fukuyama focuses on the importance of building social capital at the microlevel, a prime example of this being microfinancing strategies.

Microfinancing can be particularly useful when used for disaster risk reduction, as studies have shown that it both increases resilience and decreases vulnerability postdisaster (Marincioni et al., 2013). In terms of building the social capital of women, there are several ways in which social capital can be enhanced for women, including enhanced interaction among disaster survivors after a disaster strikes, establishing leadership programs that include women as participants, and allowing institutions to be created and thrive in a postdisaster context (i.e., civic networks) (Ganapati, 2012).

In addition, understanding how to better foster the development of networks is critical to enhance resilience. Networks are an important part of social capital, and more often than not, women that belong to richer, more diverse networks predisaster are above the poverty level and are typically not from underrepresented groups; this was especially evident in the context of Hurricane Katrina (Litt et al., 2012). Litt et al. (2012) describe two types of networks: those that are flat, thin, and homogenous, and those that are heterogeneous and diverse. Based on their casework, they argue that women with heterogeneous networks predisaster are better suited to recover postdisaster. To best foster creation of these networks and leverage this for postdisaster resilience, planners must work to bring groups together and include them in the planning processes. These groups, whether formal or informal, should have close ties with disaster management officials to best be prepared to disseminate pre-, during, and after-disaster plans; this is especially important in terms of warnings and evacuations.

Gender Equity

As previously discussed, a community's gender structure predisaster will have a great effect on the gender relations postdisaster. Women play extremely important roles in the recovery process; however, these roles are usually thrust upon them because they are similar to the work performed in the private sphere. What needs to occur is a shift in recognizing what kinds of work are more valuable than others. This can be achieved by allowing more participation by women in the disaster planning processes. It has been noted several times that the fields of emergency management and disaster-related planning are dominated by males. Thus, including a female perspective is of the utmost importance to the integration of gender-sensitive disaster policies.

Furthermore, the case studies demonstrated that the disbursement of aid could negatively affect communities. More often than not, aid is given and relief is provided with the concept that it is better to move quickly and help as many people as possible as quickly as possible with regard for the consequences that may occur from this process. This quick disbursement of aid can have the negative effect of exacerbating the existing gendered inequalities, and many women may not be in the position of those in El Hatillo to organize and understand how this aid, if managed properly, could lead to their emancipation.

Gender and Disaster Mainstreaming

The key to implementing policies that enhance the resiliency of women falls under the umbrella of gender mainstreaming. This term is defined by the United Nations Development Programme (UNDP, 2007) as "a strategy for making women's as well as men's concerns and experiences an integral dimension of... the policies and programmes in all political, economic and societal spheres so that women and men benefit equally and inequality is not perpetuated" (p. 5). To achieve this mission, the United Nations International Strategy for Disaster Reduction (UNISDR) helps support communities by developing tools and policy recommendations that can be implemented to enhance gender sensitivity. The foundations of their policy suggestions point to the need to recognize the rights of every member of society, and reflect these rights in disaster risk reduction (DRR), a focus on reorganizing society and current institutions that do not reflect gender equity, and the encouragement of participation of women in the development of policy (UNISDR, 2009, p. 28). Essentially, the UNISDR encourages a model of sustainable development in all areas of society, which will most likely require innumerable policy changes in many countries, as a gender perspective has not been applied adequately or, in many cases, taken into consideration.

As the UNISDR states, "a resilient community is a gender sensitive community"; this means that for a community to be truly resilient, every person within that community must be afforded the same opportunities to recover and bounce back. Policies must be tailored to individual communities to reduce not only gender inequities, but also racial and class inequities, as these all intersect with one another and contribute to enhanced vulnerabilities.

Improving Community Preparedness and Response

According to Pyles (2007), community leaders need to create a policy environment that encourages community organization and sustainable development to "reduce vulnerability by addressing the root causes of disasters and the lack of access to economic and political tools" (p. 322). Furthermore, Seballos and Harris (2012) explain that policies aimed at enhancing resilience must be flexible and adaptable, engage the local communities, and focus on the enhanced development of horizontal and vertical integration. The following recommendations are divided into three sections: community-based disaster management, improving vertical and horizontal integration, and postdisaster planning.

Community-Based Disaster Management

The community-based disaster management (CBDM) method attempts to correct the top-down approach that failed to meet the needs of vulnerable populations and ignored the potential of local resources and capacities (Victoria, 2001). This method of disaster management is encouraged in both developing and developed nations. This approach is widely supported due to its focus on engaging all stakeholders within the community and mobilizing the community as a whole to engage in disaster mitigation, preparation, and recovery planning. The Federal Emergency Management Agency (FEMA) contends that emergency managers, community-based organizations (CBOs), and faith-based organizations (FBOs) should work together to achieve the goal of disaster preparedness; each of these groups serves a diverse population and has access to specialized resources that can benefit the community as a whole. According to FEMA (2004), CBOs and FBOs are critical, as they "can provide a truly 'bottom-up' approach to mitigation, featuring an emphasis on social, rather than technological, solutions and empowerment of the local community" (p. 8).

The push for CBDM needs to start with support from the government. Without this support, any projects developed may not receive the adequate funding to reach fruition. The first step is to develop strong partnerships between the local emergency managers or government officials and the appropriate CBOs and FBOs. These partnerships should be based on the mutual interest of enhancing community preparedness and resilience; thus, all parties must be committed to meeting and establishing the needs and goals of such a partnership. Each group should bring in ideas from its respective stakeholders; focus groups or community outreach should be completed to canvas all applicable needs so that they are adequately represented. Once these partnerships are established, an action plan must be developed. This action plan is the basis for any mitigation projects that will be established within the community (FEMA, 2004, pp. 3–4). Action plans are enhanced and informed by adequate risk assessments and a complete inventory of the resources that the community already has available.

In addition to the encouragement of FEMA to adopt a CBDM approach in the United States, organizations such as the World Bank also recognize the importance of this strategy. The World Bank's 2009 approach was focused on the development of social funds and community-driven development (CDD), defined as funds provided by "government agencies or programs that channel grants to communities for small-scale development projects" (i.e., infrastructure, investments, social services) (p. 6). There are several advantages to this social fund/CDD approach, as these are programs that exist in many countries and possess both a local and a national presence; they have demonstrated that they are efficient, effective, and "have a good track record for incorporating lessons learned into longer-term development strategies" (p. 8). Thus, examining the effective integration of these programs in developing countries could substantially reduce their vulnerability and enhance not only their disaster resilience capacity but also their community as a whole.

Overall, there are myriad resources available to emergency managers, government officials, and leaders of local community organizations that explain step-by-step how to create and maintain an effective CBDM system. It is critical that all communities embrace this approach and recognize where improvements can be made to enhance their resiliency.

Improving Vertical and Horizontal Integration

Vertical and horizontal integration are of the utmost importance when ensuring speedy and efficient recovery after a disaster. The framework of horizontal and vertical integration is important to understand, as when both of these are operating at high capacity, communities are able to successfully navigate their local needs and the hierarchy of local, state, and federal organizations to get the help they need after a disaster (Smith et al., 2013). This concept is highly relevant to the CBDM strategy, as those communities with high integration will be more cohesive and more likely to mobilize and be resilient in the postdisaster context.

To better integrate these dimensions within a community, Berke et al. (1993) made several suggestions. The first reflects the increased involvement of the community and local government. As previously discussed, these partnerships are extremely important to establish resilience. Engaging the community increases horizontal integration because it builds upon the capacities of individuals, organizations, and the government, leading to a more cohesive front in the face of a disaster. By focusing on CDBM, horizontal integration can be significantly strengthened. Furthermore, Berke et al. (1993) state that the strengthening of this dimension has a positive influence on enhancing vertical integration as well, since "the experiences documented by the limited research on disaster recovery shows that vertical integration can be more effective at meeting local needs when activities that strengthen horizontal integration before and during recovery are present" (p. 106).

Postdisaster Planning

In addition to mitigation and preparedness strategies, communities can increase their resilience through engaging in the development of postdisaster planning. These redevelopment plans acknowledge that their community could be at risk, encouraging stakeholders to identify their vulnerabilities before they are exacerbated by an event. Berke et al. (1999) encourage communities to use a disaster recovery tool typology to determine how well their redevelopment plans are structured. This typology includes the following:

> (1) Regulatory measures (e.g., building codes, zoning, development moratorium), which are coercive in that they attempt to control the activity of specific interest groups;
> (2) incentive measures (e.g., development density bonus, capital improvement program, property acquisition), which are noncoercive in that they aim to induce, rather than require, desired redevelopment; and
> (3) informational measures (e.g., rebuilding workshops, reconstruction plans, dissemination programs on availability of disaster assistance), which enable people to make informed redevelopment decisions (p. 106).

By examining the presence of these measures, communities can determine how much work needs to be done to ensure that they are sufficiently prepared to recover efficiently and successfully from a disaster.

An additional strategy that can be used to understand and address the vulnerabilities of a community to best enhance postdisaster resilience is to construct vulnerability maps. These maps are specific to the social and political context of a community, and help planners to understand how to tailor mitigation plans. Planners must keep in mind the concept of intersectionality when analyzing areas of vulnerability within the community; the best mitigation strategies will include those plans that look at not only social and political contexts, but also how these contexts interact with gender, economic status, and cultural practices (Morrow, 1999).

It is clear that each of these recommendations can be implemented in a wide variety of contexts to enhance the resilience of both women and their communities. One of the key themes throughout this chapter is the concept of strong community cohesion. Those communities that are most cohesive also tend to have higher levels of predisaster gender equity and strong horizontal and vertical integration. It is imperative to engage all stakeholders in the disaster planning process, as the resilience of the individuals in the community will have a great impact on the resilience of the community as a whole.

CONCLUSION

Although there is evidence that gender-based vulnerabilities have an effect on the resilience of a community postdisaster, there is still a significant amount of work to do in identifying policies that will resolve this problem in both developed and developing countries. The chapter provided a brief review of the literature and expanded on these findings through several case studies. The lessons drawn from both the literature and case studies provide guidance for practical policy development. It is important to note that these guidelines should be implemented according to the level of inequality or lack of cohesion present in a given community. There is not a one-size-fits-all approach to better preparing communities for disaster; thus, a community inventory must be completed and all stakeholders within the community should be involved. A cohesive community that values gender equity is imperative to ensure that it is resilient after a disaster. Policy makers and local government officials need to recognize the importance of developing solid disaster mitigation, preparedness, and recovery plans to guarantee that their communities will recover and thrive should they be struck by a disaster.

REFERENCES

Berger, A., and Krahé, B. (2013). Negative attributes are gendered too: Conceptualizing and measuring positive and negative facets of sex-role identity. *European Journal of Social Psychology*, 43(6), 516–531.

Berke, P. R., Kartez, J., and Wenger, D. (1993). Recovery after disaster: Achieving sustainable development, mitigation and equity. *Disasters*, 17(2), 93–109.

Bruneau, M., Chang, S. E., Eguchi, R. T., Lee, G. C., O'Rourke, T. D., Reinhorn, A. M., and von Winterfeldt, D. (2003). A framework to quantitatively assess and enhance the seismic resilience of communities. *Earthquake Spectra*, 19(4), 733–752.

Cram, B. (2014). Women in the face of disaster: Incorporating gender perspectives into disaster policy. Quick Response Grant Report series, 247.

Cupples, J. (2007). Gender and Hurricane Mitch: Reconstructing subjectivities after disaster. *Disasters*, 31(2), 155–175.

Cutter, S. L., Barnes, L., Berry, M., Burton, C., Evans, E., Tate, E., and Webb, J. (2008). A place-based model for understanding community resilience to natural disasters. *Global Environmental Change*, 18(4), 598–606.

Cutter, S. L., Boruff, B. J., and Shirley, W. L. (2003). Social vulnerability to environmental hazards. *Social Science Quarterly*, 84(2), 242–261.

David, E. (2010). "Studying up" on women and disaster: An elite sustained women's group following Hurricane Katrina. *International Journal of Mass Emergencies and Disasters*, 28(2), 246–269.

Donner, W., and Rodríguez, H. (2008). Population composition, migration and inequality: The influence of demographic changes on disaster risk and vulnerability. *Social Forces*, 87(2), 1089–1114.

Enarson, E. (2012). *Women Confronting Natural Disaster: From Vulnerability to Resilience*. Boulder, CO: Lynne Rienner Publishers.

Enarson, E., Fothergill, A., and Peek, L. (2007). Gender and disaster: Foundations and directions. In *Handbook of Disaster Research*, 130–146. New York: Springer.

FEMA. (2004, January 1). Community-based pre-disaster mitigation for emergency managers. Retrieved April 25, 2014, from http://www.fema.gov/media-library-data/20130726-1445-20490-5821/resourceguide.pdf.

Fothergill, A. (1999). Women's roles in a disaster. *Applied Behavioral Science Review*, 7(2), 125–143.

Fothergill, A. (2003). The stigma of charity. *Sociological Quarterly*, 44(4), 659–680.

Fukuyama, F. (2002). Social capital and development: The coming agenda. *SAIS Review*, 22(1), 23–37.

Ganapati, N. E. (2012). In good company: Why social capital matters for women during disaster recovery. *Public Administration Review*, 72(3), 419–427.

Ganapati, N. E. (2013). Downsides of social capital for women during disaster recovery toward a more critical approach. *Administration & Society*, 45(1), 72–96.

Laska, S., Morrow, B. M., Willinger, B., and Mock, N. (2008). Gender and disasters: Theoretical considerations. In *Katrina and the Women of New Orleans*, ed. B. Willinger, 11–21. New Orleans, LA: Tulane University.

Litt, J., Skinner, A., and Robinson, K. (2012). The Katrina difference: African American women's networks and poverty in New Orleans after Katrina. In *The Women of Katrina: How Gender, Race, and Class Matter in an American Disaster*, ed. E. David and E. Enarson, 15–23. Nashville, TN: Vanderbilt University Press.

Marincioni, F., Appiotti, F., Pusceddu, A., and Byrne, K. (2013). Enhancing resistance and resilience to disasters with microfinance: Parallels with ecological trophic systems. *International Journal of Disaster Risk Reduction*, 4, 52–62.

Morrow, B. H. (1999). Identifying and mapping community vulnerability. *Disasters*, 23(1), 1–18.

Paton, D. (2006). Disaster resilience: Building capacity to co-exist with natural hazards and their consequences. In *Disaster Resilience: An Integrated Approach*, ed. D. Paton and D. M. Johnston, 3–10. Springfield, IL: Charles C. Thomas.

Plodinec, M. J. (2009). *Definitions of Resilience: An Analysis*. Washington, DC: Community and Regional Resilience Institute.

Plough, A., Fielding, J. E., Chandra, A., Williams, M., Eisenman, D., Wells, K. B., and Magaña, A.. (2013). Building community disaster resilience: Perspectives from a large urban county department of public health. *American Journal of Public Health*, 103(7), 1190–1197.

Putnam, R. D. (1993). The prosperous community: Social capital and public life. *American Prospect*, 13, 35–42.

Pyles, L. (2007). Community organizing for post-disaster social development: Locating social work. *International Social Work*, 50(3), 321–333.

Pyles, L., and Lewis, J. S. (2007). Women of the storm: Advocacy and organizing in post-Katrina New Orleans. *Affilia*, 22(4), 385–389.

Rose, A. (2007). Economic resilience to natural and man-made disasters: Multidisciplinary origins and contextual dimensions. *Environmental Hazards*, 7, 383–398.

Seballos, F., and Harris, K. (2012). Re-shaping policy and institutions for integrating climate and disaster resilience. Retrieved from http://www.preventionweb.net/english/professional/publications/v.php?id = 24913.

Smith, G., Sandler, D., and Goralnik, M. (2012). Assessing state policy linking disaster recovery, smart growth, and resilience in Vermont following tropical storm Irene. *Vermont Journal of Environmental Law*, 15, 66–102.

Symington, A. (2004). Intersectionality: A tool for gender and economic justice. *Women's Rights and Economic Change*, 9, 1–8.

Tierney, K. (2006). Social inequality, hazards, and disasters. In *On Risk and Disaster: Lessons from Hurricane Katrina*, ed. R. Daniels, D. Kettl, and I. Kunreuther, 109–128. Philadelphia, PA: University of Pennsylvania Press.

Tobin, G. A. (1999). Sustainability and community resilience: The holy grail of hazards planning? *Global Environmental Change Part B: Environmental Hazards*, 1(1), 13–25.

United Nations Development Programme, Environment and Energy Group. (2007). Gender mainstreaming: A key driver of development in environment and energy. Retrieved from http://www.undp.org/content/dam/undp/library/Environment and Energy/Sustainable Energy/Gender_Mainstreaming_Training_Manual_2007.pdf.

United Nations. (2005). Hyogo Framework for Action 2005–2015: Building the resilience of nations and communities to disasters. Retrieved from http://www.unisdr.org/2005/wcdr/intergover/official-doc/L-docs/Hyogo-framework-for-action-english.pdf.

United Nations International Strategy for Disaster Reduction. (2009). Making disaster risk reduction gender-sensitive: Policy and practical guidelines. Retrieved from http://www.unisdr.org/files/9922_MakingDisasterRiskReductionGenderSe.pdf.

United Nations International Strategy for Disaster Reduction. (2011). Hyogo Framework for Action 2005–2015: Building the resilience of nations and communities to disasters mid-term review 2010–2011. Retrieved from http://www.preventionweb.net/files/18197_midterm.pdf.

Victoria, L. P. (2001). Community-based Approaches to Disaster Mitigation. Available online at http://www.unpan1.un.org/intradoc/groups/public/documents/APCITY/UNPAN009661.pdf

What is community resilience? (2014). Community and Regional Resilience Institute (CARRI). Retrieved April 25, 2014, from http://www.resilientus.org/about-us/what-is-community-resilience/.

What is social capital? (2007). *OECD Insights: Human Capital*. Retrieved April 25, 2014, from http://www.oecd.org/insights/37966934.pdf.

World Bank. (2009, January 1). Building resilient communities. Retrieved April 25, 2014, from http://siteresources.worldbank.org/INTSF/Resource s/Building_Resilient_Communities_Complete.pdf.

2

Nonprofits and Disasters

Grace L. Chikoto

Contents

INTRODUCTION

There are reports of an unprecedented dramatic upward trend in the occurrence of natural catastrophes worldwide, largely attributed to climate change, with a number of geophysical (earthquakes, tsunami), meteorological (storms), hydrological (floods, mass movements), and climatological (extreme temperature, forest fires) events, increasing from under 400 events in 1980 to nearly 900 events in 2012 (Munich Re, 2012). Specific to the United States, the Federal Emergency Management Agency (FEMA, 2008) foresees a "high vulnerability to natural hazards, weapons of mass destruction, and widespread disease epidemics" among U.S. urban cities (World Bank, 2010). Already disasters like the 9/11 terrorist attacks, Hurricanes Rita and Katrina, Superstorm Sandy, and the recent 2013 Oklahoma tornadoes all demonstrate the scale and trail of destruction and devastation disasters have on various dimensions of a community.

One dimension also at stake comprises the diverse nonprofit organizations that serve various population groups in our communities. Post-Katrina reports and surveys clearly demonstrate the devastating impact natural disasters have on nonprofit organizations (Auer and Lampkin, 2006; Vita and Morley, 2007; Weisbrod and Asch, 2010). On the other hand, the frequency of natural disasters seeks to remind us of the existence of a physical and natural system that is largely influenced by both physical laws and human interference, thus sounding a call for disaster preparedness (Haimes, 2012). From the vantage point of the impact of Hurricanes Katrina and Sandy, the costliest hurricanes recorded since 1980 (Munich Re, 2014), this chapter considers the state of disaster mitigation and preparedness among nonprofit organizations.

This chapter is primarily motivated by two factors: the unprecedented frequency of disasters and the widespread functions and roles the nonprofit sector plays within the U.S. economy and society, including preparing communities for disasters and responding to disasters. First, the nonprofit sector came of age as a formidable economic and social force in the U.S. economy and welfare system (Salamon, 2003; Smith, 2006), playing key supplementary and complementary roles (Young, 2000, 2006a). The importance of nonprofits is also exemplified by a growing nonprofit–government partnership (Smith and Lipsky, 1993; Salamon, 2003) that is, in part, a consequence of the hollowing of the state (Milward and Provan, 2000), and one that has culminated into a widespread disengagement by the state from direct service provision (Salamon, 1995).

Second, related to disaster preparedness and response, most of the research focus was on nonprofit organizations' *response* roles in the aftermath of disasters, with a few insights on their own state of *mitigation and preparedness* efforts (Chikoto et al., 2013). This is in spite of the fact that nonprofits historically played significant roles in the nation's disaster response system, both formally and informally. Emerging from postdisaster tales were unmistakable praises of nonprofit organizations for being among crucial first responders in their agility in responding to disaster victims (Gajewski et al., 2010; Kapucu, 2006b) relative to the government (Kapucu and Wart, 2006; Green et al., 2007). In particular, nonprofits' potency was largely heightened by the important roles they played in response to the 9/11 terrorist attacks and Hurricanes Katrina and Rita (Fagnoni, 2006).

Note that the nonprofits referenced in this chapter differ from National Voluntary Organizations Active in Disaster (NVOAD)* members, such as the American Red Cross, the Billy Graham Rapid Response Team, and the Brethren Disaster Ministries. Instead, this chapter is mostly concerned with nonprofits that serve vulnerable populations, in particular social and human services nonprofits (e.g., youth, spouse, or child violence prevention, food pantries and soup kitchens, homeless shelters, etc.) that are "thrust into or voluntarily" (Simo and Bies, 2007) play unanticipated disaster response roles in the aftermath of disasters.

THE FOUR STAGES OF EMERGENCY MANAGEMENT

Developed by the National Governors' Association in the 1970s, the emergency management model identifies four phases or processes: disaster *mitigation*, *preparedness*, *response*, and *recovery* (Waugh, 1999). Although simplistic, the four phases identify functional categories that facilitate disaster administration (Waugh, 1999). As others have noted, collaborative emergency management requires horizontal and networked partnerships with all segments of society, especially public, private, *and* nonprofit sectors in order to be comprehensive and successful (Demiroz and Hu, 2014; Comfort et al., 2012; FEMA, 2011; Kapucu, 2006a). In addition, the National Disaster Recovery Framework (NDRF) also recognizes that nonprofit involvement in emergency management comes from a plethora of organizations, ranging from "small locally-based nonprofits to national organizations with extensive experience in disaster recovery" (FEMA, 2011). Hence, nonprofit organizations are crucial partners in disaster preparedness, mitigation, response, and recovery (Demiroz and Hu, 2014), whether or not disaster management is their primary mission.

* For a discussion of NVOAD, see Kapucu, N., Yuldashev, F., and Feldheim, M.A., "Nonprofit Organizations in Disaster Response and Management: A Network Analysis," *European Journal of Economic and Political Studies*, 4(1), 83–112, 2011.

Mitigation consists of predisaster activities involving risk assessment and the implementation of measures to minimize the potential effects of disaster (Waugh, 1999). Mitigation strategies include land use regulations, structural barriers to prevent or control hazards, building codes, and purchasing an insurance program to minimize the economic impact of disasters. Generally, such measures aim to control hazard sources in order to reduce impact. For example, building levees, fortifying property or structures, retrofitting parts of a building, strapping water heaters, controlling sources of hazards, adopting land use practices, and protecting the contents of a building are all measures that serve to reduce the extent of damage and do not require activation at the time of disaster. Hence, they passively provide protection to persons and property at the time of disaster impact (Lindell and Whitney, 2000; Arlikatti et al., 2007).

Also predisaster strategy, the process of disaster *preparedness* entails getting ready for expected threats (Waugh, 1999). Activities include engaging in contingency planning, resource management, and mutual and cooperative agreements with external partners, providing public information, and training response personnel (Waugh, 1999). Within the preparedness phase, however, some draw distinctions between *response* and *recovery* activities. *Response preparedness* strategies (e.g., incident management actions, staff drills, and training) are designed to actively equip actors and organizations with responsive capacity at impact and therefore "support active response when a disaster strikes" (Arlikatti et al., 2007). In contrast, *recovery preparedness* measures, such as flood insurance policies, merely provide support with the recovery process in the aftermath of a disaster (Lindell and Perry, 2000). Such measures do not provide organizations with a mechanism to escape or reduce the likelihood of loss from occurring in the first place (Herman, 2004). To illustrate, Hurricane Katrina resulted in an approximate $125 billion in overall property damage and loss, of which only $62.2 billion was insured (Munich Re, 2012). Mitigation and preparedness are therefore key strategies that *all* nonprofits ought to consider and implement.

Response activities, on the other hand, are strategies implemented *during* a disaster. Such activities include evacuations, search and rescue, providing emergency medical services, and firefighting, to name a few (Waugh, 1999).

Finally, the fourth phase of disaster *recovery* begins in the first few hours of the disaster and may last decades (Demiroz and Hu, 2014). Strategies in this phase are postdisaster activities designed to restore basic services, such as repairing water and telecommunication systems and power lines (Waugh, 1999). Additional recovery activities include providing temporary housing, food, and clothing; providing counseling services; and assisting victims with job searches and loans for restarting businesses.

Recovery also entails long-term community reconstruction, with an aim to fortify the community for future disasters. Hence, the process "determines whether a community can bounce back to pre-disaster conditions or develop stronger community capacity," that is, the disaster resilience of a community (Demiroz and Hu, 2014). In all, disaster recovery is "a complex and long lasting practice" that involves numerous cross-sector actors from "public, private, and nonprofit sectors" (Demiroz and Hu, 2014). Hence, similar to disaster response, recovery is a collaborative emergency management effort involving nonprofits, community-based and civil society organizations, public agencies, and private organizations.

This chapter provides a review of the current demographics of the nonprofit sector displaying the keys roles they play in U.S. communities. This is followed by a portrayal of nonprofits' encounter with two key natural disasters in order to demonstrate how the 2005 Hurricane Katrina and the 2012 Hurricane Sandy disasters impacted nonprofits and how nonprofits responded to these disasters. The chapter goes on to give a brief overview of the state of the literature on nonprofit emergency management, with an emphasis on disaster mitigation and preparedness. The chapter concludes by addressing some of the challenges and opportunities that nonprofit organizations face with respect to implementing disaster mitigation and preparedness strategies.

In summary, nonprofits, in particular, human service nonprofits, have the advantage of being embedded in local neighborhoods and communities, in addition to being able to directly supplement and fill gaps where government falls short. Hence, beyond NVOAD, this chapter focuses on those nonprofits that are often thrust into disaster response and recovery roles regardless of their nondisaster missions. As Waugh and Streib (2006) pointed out, "emergency management capacity is built from the ground up," and hence "neighborhoods and communities programs have to be able to stand on their own" until external assistance arrives, which may take hours or days. As such, nonprofit organizations as a type of local social capital have a part to play in emergency management, in addition to ensuring that they are at least prepared for disasters themselves.

U.S. NONPROFIT ORGANIZATIONS: DIVERSITY AND SIGNIFICANCE

Nonprofit organizations (also known as public charities) in the United States are generally distinguished from other organizational forms (e.g., public agencies and private corporations) by six key factors: tax exemption status, nondistribution constraint, organizationally separate from government, voluntary, public/mutual benefit, and self-governance (Hammack, 2002; Salamon, 2003). Section 501(c) of the Internal Revenue Service (IRS) code defines the nonprofit organizational form as a type of organization that is precluded from paying a federal income tax on its revenue (tax exemption status). While this tax exemption status applies to more than 30 types or classifications of nonprofit organizations that provide mutual and public benefits (Salamon et al., 1999), only nonprofits that exist for the public benefit are permitted to receive tax-deductible contributions or donations under 501(c)(3) of the IRS code (Blackwood et al., 2012; Brody, 2006). Of the 2.3 million nonprofits that operated in the United States in 2010, approximately 1 million of these were 501(c)(3) organizations (Blackwood et al., 2012). Of the 1 million 501(c)(3)s that filed a 990 form with the IRS, 34% comprised human service* nonprofits—the largest service field among public charities (Blackwood et al., 2012).

Similar to public agencies, nonprofits are prohibited from private inurement, that is, from distributing any profit to their stakeholders or board of directors (Hansmann, 1980). In the spirit of self-governance, the boards of directors referenced here are self-perpetuating boards comprised of volunteers who govern and provide direction for the organization in its attempts to fulfill its mission. Additionally, the board also exists to execute fiduciary duties of loyalty and care by elevating and safeguarding the interests of the organization (Brody, 2006). As such, nonprofit organizations are separate from government.

As demonstrated by voluntary boards of directors, volunteerism (in terms of time and money) is an integral part of the financial makeup of the nonprofit sector, with approximately 27% of the U.S. population (64.3 million

* Generally, human services nonprofits include those involved in crime- and legal-related activities (e.g., youth violence prevention; spouse, child, and sexual abuse prevention; inmate support); employment (e.g., vocational counseling and job training); food, agriculture, and nutrition, but with a primary focus on food programs (e.g., food banks and pantries, soup kitchens and meals on wheels); youth development (e.g., Boy Scouts of America, Big Brothers Big Sisters, and various youth development programs); and multipurpose human service organizations such as adult care, homeless centers, family counseling, and child care.

people, equivalent to 8.9 million full-time employees worth $296.2 billion) reporting volunteering at least once in 2011 (Blackwood et al., 2012). Using U.S. Census data, the Corporation for National and Community Service (2014) estimated that one in four Americans volunteers.

Nonprofits are also distinguished by their reliance on diverse funding streams (Blackwood et al., 2012; Young, 2006b), with 21% coming in the form of cash and in-kind donations, and foundation and government grants; 73% from fees for service, which often include government contracts; and 6% from other sources of income, such as dues, rental income, special event income, and gains or losses from goods sold (NCCS, 2012).

Also noteworthy is that nonprofits often encounter unique challenges in mobilizing the necessary resources to support and sustain their operations and activities, as well as donor-imposed restrictions on how funds can be utilized (Young, 2006b). This is especially so for smaller nonprofits. For instance, in terms of organizational size (measured by total expenses), the nonprofit sector is comprised of very small organizations, with 74% of nonprofits spending under $250,000 annually (Blackwood et al., 2012) and 45% of all reporting nonprofits spending less than $100,000 annually. This is a crucial issue, as risk management planning demands temporal, financial, and human resources (Herman, 2004; Herman et al., 2003; Young, 2003). Research shows a negative correlation between fiscal and human resources and investments in disaster preparedness (Lindell and Perry, 2000) and continuity planning (Meyer-Emerick and Momen, 2003).

Nonprofit Roles: The Diversity of Sector

In terms of the scope of the sector, even when relying on volunteers, nonprofits remain a significant employment contributor, registering a 17% growth between 2000 and 2010 (Blackwood et al., 2012). In addition to their 5.5% share of the U.S. gross domestic product (GDP), nonprofits accounted for 9.2% of all wages and salaries in 2010 (Blackwood et al., 2012).

Furthermore, nonprofits serve as *gap fillers* (Anheier, 2005), playing key supplementary and complementary service provision roles (Young, 2000, 2006a; Salamon, 2003), often to correct government and market failures (Steinberg, 2006). In particular, nonprofit roles are distinguished from their private and public sector counterparts in four key ways: the vanguard roles they play in generating innovative ideas; as value guardians that help foster the expression of diverse values; in their advocacy roles, giving voice to the underrepresented and shaping public policy; and as service providers (Kramer, 1982), as articulated above. To these four important contributions distinguishing nonprofits, Salamon et al. (2000) added the role of community building and democratization.

Data from the National Center for Charitable Statistics (NCCS) portrays the contributions nonprofits make using the National Taxonomy of Exempt Entities (NTEE) classification. Based on the NTEE classification, nonprofits play significant roles in the arts, culture, and humanities (*N* = 39,536); education, higher education and otherwise (*N* = 66,769); environment (N = 16,383); health, including hospitals and other health care facilities (*N* = 44,128); public and societal benefit (*N* = 43,875); religion (*N* = 23,502); and human services (*N* = 124,360) (Blackwood et al., 2012). These numbers are based on the 366,086 nonprofits that filed a 990 form with the IRS in 2010, suggesting incomplete polling of all nonprofits and civic associations embedded in our communities (Boris and Steuerle, 2006; Smith, 1997). Note that nonprofits not polled also included religious congregations and nonprofits with less than $5,000 in annual total revenues, as they are not required to register with the IRS, let alone report their finances (Blackwood et al., 2012; Boris and Steuerle, 2006).

As alluded to earlier, another dimension for gauging the significant functions of the nonprofit sectors is exemplified in the prevalent government–nonprofit partnership that first took root in the second half of the twentieth century (Smith and Lipsky, 1993; Smith, 2006; Toepler, 2010). The rise in government financing of nonprofit activities post–World War II resulted in the nonprofit sector's growth from a small cottage industry of nearly 12,500 charities in the 1940s (Hall, 2004) to the formidable economic force it is today (Salamon et al., 1999; Toepler, 2010). In fact, such state–nonprofit partnerships allowed the carrying out of public purposes through private provision by nonprofit organizations (also see Smith and Lipsky, 1993; Salamon, 1995; Smith, 2006). The introduction of Charitable Choice in the mid-1990s and the subsequent establishment of the White House's Faith-Based Initiative in 2001 further highlight the essential nature of the nonprofit sector in correcting various social ills.

In light of the sheer size, scope, and contributions of this important form of social capital, combined with increased occurrences of natural disasters and organizational vulnerability to natural disasters, an inquiry into nonprofits' state of disaster planning and management is warranted.

NONPROFITS' ENCOUNTER WITH DISASTERS: IMPACT AND RESPONSE

While the question of the state of disaster mitigation and preparedness among U.S. nonprofits remains to be answered comprehensively, much is said about the impact natural disasters have on nonprofits, as well as the roles that are thrust upon them or their voluntary response to disasters. For one, "disasters highlight elements of community vulnerability and resiliency" (Green et al., 2007). By gleaning text-based data for information about the impact of major disasters that hit the United States in the most recent past, this section exhumes and highlights the vulnerabilities inherent among nonprofit organizations. This section attempts to paint a picture of the nature of devastation that befalls nonprofits by tracing the impact of two historic natural disasters on nonprofits: Hurricanes Katrina (2005) and Sandy (2012). This effort also highlights some of the roles some nonprofits played in response to those affected by each disaster.

Hurricane Katrina resulted in more than 1,300 deaths, with damage to hundreds of thousands of buildings and homes, and an overall loss of approximately $125 billion, of which only $62.2 billion was insured (Munich Re, 2014).

However, not all areas experienced Hurricane Katrina in the same way. Relative to southeast Louisiana, nonprofit organizations in the Mississippi Gulf Coast, in particular, experienced significant structural damage to physical and organizational infrastructures, resulting in damage to buildings, equipment, communication systems, supplies, and a loss of records (Green et al., 2007; Weisbrod and Asch, 2010). Not only did this displace service provision in New Orleans, but it also made it difficult for those needing assistance to locate service providers and for the displaced nonprofits to locate their regular clients (Green et al., 2007). For colleges and universities like Dillard University, Loyola University–New Orleans, and Tulane University, physical damage to structures compelled temporary closures (Weisbrod and Asch, 2010) and relocation for others. Tulane University, for instance, was forced to relocate to Houston for the 2006–2007 academic year, resulting in significant student dropouts (Taylor et al., 2006).

Before Hurricane Katrina, there were 3,200 reporting nonprofits (filing a 990 form with the IRS) in the state of Louisiana, with nearly 900 located in New Orleans; this represented nearly half of Louisiana-based nonprofit expenditures and assets (Urban Institute, 2005). Of these, 385 nonprofits provided human services and community improvement programs to New Orleans residents (Urban Institute, 2005).

Post–Hurricane Katrina, however, a Louisiana Association of Nonprofit Organizations (LANO) survey reported that some 95% of the 212 health and human services nonprofits surveyed reported being directly or indirectly impacted by Katrina. Auer and Lampkin (2006) suspected that some 98 of these nonprofits temporarily or permanently closed. In the Biloxi–Pascagoula metro area, 67% of the area nonprofits suffered losses to paid and volunteer staff, with 77% reporting major losses or damage to buildings, and 93% of those surveyed by the Mississippi Center for Nonprofits reporting losses to programs and services (Pipa, 2006).

From a funding perspective, while some nonprofits attracted sharp increases in donations, including grants to expand service provision following the hurricane, many experienced cuts in government funding (Green et al., 2007). Colleges and universities encountered post-Katrina drops in total revenues between 2005 and 2006. For example, Tulane University experienced a 10% revenue loss during this period, including a 45% drop in university-related hospital revenues (Weisbrod and Asch, 2010).

With respect to nonprofits' responses to Hurricane Katrina, it is important to recognize that of those nonprofits that did respond, for many, disaster relief or recovery was not their *raison d'etre*. In other words, there were non-NVOAD members. Nonetheless, with limited experience and resources, respond they did. Overall, nonprofits (including faith-based organizations) responded to human needs by supplying food and water, through "meals ready to eat" (MREs), soup kitchens, and food pantries (Green et al., 2007). Additional assistance included collecting and distributing clothing, personal hygiene products, and other general household items; providing emergency shelters, health care, and home cleaning services; and providing advocacy services for the affected poor minority and ethnic groups, and for vulnerable populations such as children and the elderly (Green et al., 2007).

In terms of organizational lessons, common themes emerged from the nonprofits' staff interviewed by Green et al. (2007), including the need for nonprofits to rely less on government assistance, their need to be prepared to confront disasters on their own, and the essentiality of adopting "disaster plans for their own service delivery efforts."

Aside from Hurricane Katrina, Hurricane Sandy was the second costliest disaster since 1980 (Munich Re, 2014). Affecting parts of the United States, the Bahamas, Cuba, the Dominican Republic, Haiti, Jamaica, Puerto Rico, and Canada, Hurricane Sandy resulted in 210 deaths and an overall loss of approximately \$68.5 billion, of which \$29.5 billion was insured (Munich Re, 2014).

Relative to other New York City boroughs, Staten Island's vibrant nonprofit sector historically struggled to attract off-island funding despite having comparatively fewer public services, greater public health needs due to the highest all-cause mortality rates in the area, and the highest youth alcohol and substance abuses in New York City (Major, 2013). Based on a Human Services Council of New York survey conducted following Superstorm Sandy, two-thirds of the 382 nonprofits that responded to the survey experienced temporary closures (NYNP, 2012). Together, these nonprofits were estimated to have served some 333,511 individuals, offering services such as mental health support, after-school programs, youth services, community prevention programs, and food provision before Hurricane Sandy (NYNP, 2012). However, as a result of the hurricane, these nonprofits were unable to serve 144,318 clients (NYNP, 2012).

Based on a pre–town hall survey conducted by the Community Resource Exchange (CRE) on the 56 nonprofit town hall participants, only 11 of these were not impacted by Hurricane Sandy (CRE, 2013). Thirty-five nonprofits experienced closures or interruption to services ranging from several days ($N = 16$) to more than 3 weeks ($N = 5$) (CRE, 2013). The qualitative portions of the survey further revealed the extent of the impact. Some nonprofits experienced closures in after-school and child care programs, with some child care facilities and property experiencing substantial damage. Other damages included flooding amid interruptions to telecommunications, heating, gas, and electric services (some senior centers closed as a result) (CRE, 2013).

From a funding perspective, nonprofits dependent on program service revenue saw declines in total revenue. For instance, Snug Harbor incurred significant losses due to tour cancellations and declines in the number of school trips and rentals, which in turn impacted its operations (CRE, 2013). For another organization, the impact of the hurricane resulted in a financial loss of approximately $300,000 due to nonreimbursement services since the nonprofit could neither reach its clients nor provide on-site services such as crisis counseling (CRE, 2013). Another experienced a decline in the number of clients able to take the classes the nonprofit offered, thus compromising the organization's revenue base and its ability to offer scholarships (CRE, 2013). Yet another nonprofit lost the opportunity to earn revenue, as its future site of a national lighthouse museum sustained $200,000 worth of damages (CRE, 2013).

Although the above represented a small number of vignettes from the few nonprofits interviewed, these organizations represent many others. Hence, behind these numbers lie segments of society and communities whose livelihood were adversely impacted, as they depended on one nonprofit or another.

NONPROFIT DISASTER MITIGATION AND PREPAREDNESS: GLEANINGS FROM THE LITERATURE

Within the context of nonprofit disaster management, research is largely centered on the roles nonprofits play (Gajewski et al., 2010; Simo and Bies, 2007), cross-sector collaborations (Kapucu and Wart, 2006; Simo and Bies, 2007; Syles, 2008–2009; Waugh and Streib, 2006), volunteerism (Rotolo and Berg, 2010), and nonprofits' responses to emergent needs (Auer and Lampkin, 2006; Vita and Morley, 2007). As a result, a specialized focus on disaster or emergency management strategies among nonprofits is needed, since the nonprofit literature largely favors a financial risk management focus (Young, 2003), with limited attention to disaster planning (Chikoto et al., 2013). In recognition of the mutable and uncertain environments in which these organizations operate, effective risk management becomes fundamental to the proper functioning of any organization (Young, 2003). This is especially important for nonprofits that are not NVOAD members or those that do not have disaster-related missions. As noted, nonprofits are often thrust into disaster response (Simo and Bies, 2007). Below are a few risk/emergency management themes that are addressed in the nonprofit literature.

Risk Management

Within the nonprofit context, risk management is often linked to nonprofit governance, which stresses a normative model of asset stewardship inclined toward the prevention of unnecessary erosion of assets (Brody, 2006; Herman, 2004; Herman et al., 2003; Young, 2003). This encompasses a plethora of risks associated with property; income; liability; employment practices; organizational goodwill or reputation; volunteer management; governance and fiduciary; serving vulnerable populations such as children, dependent adults, and individuals with disabilities; transporting clients; collaborations; and natural disasters (Herman et al., 2003).

Pointing to the fact that disasters are just another risk management area nonprofits managers are concerned about, Spillan (2003) found that the nonprofit managers he interviewed were more likely to be concerned about five particular vulnerability areas or crises, after controlling for disaster experience. Specifically, the nonprofit managers were more likely to be worried about their ability to remain operational in the event of major product or service equipment malfunctions or breakdowns,

breakdowns in computer systems, and theft or disappearance of organizational records; publicity related to negative media coverage, boycotts by consumers, and product sabotage; fraud as it pertained to theft of company property, embezzlement, corporate espionage, and management corruption; natural disasters, particularly floods, hurricanes, tornadoes, snowstorms; and legal issues concerning product recall, government investigation, and employee lawsuit.

The above appears to be a comprehensive coverage of nonprofit risk management; however, Herman et al. (2003) addressed natural disasters only in relation to property damage, whereas Spillan (2003) merely informed us of the types of natural disasters about which nonprofit managers were most likely to be concerned. However, as will be addressed below and in the examples from nonprofits' encounters with Hurricanes Katrina and Sandy, the impact of natural disasters is not limited only to property damage.

Organizational Vulnerability

Depending on their scale and magnitude, disasters trigger domino effects on multiple dimensions of an organization (Haimes, 2012). This fact is best demonstrated in Zhang et al.'s (2009) business vulnerability framework, and as observed in nonprofits' encounter with Hurricanes Katrina and Sandy, organizations have four key vulnerability areas: capital, in terms of incurring damage to fixed assets (buildings, equipment, and inventory such as materials and supplies); labor, whereby organizations experience temporary and permanent losses of employees and unpaid volunteer staff; suppliers or shortage of materials due to permanent relocation or closures; and customers or clients, when disasters result in short-term population dislocation or customer losses, demographic shifts, and changes in customer or client preferences.

Based on this, nonprofits are susceptible to a plethora of risks. Nonprofits have limited to zero control over the magnitude of disasters, and the recommendation is that organizations adopt predisaster actions to minimize disaster impact through the kinds of mitigation and preparedness measures they adopt (Dahlhamer and D'Souza, 1997). This underscores two issues. First, in light of the unique constraints nonprofits encounter, nonprofit managers are likely to regard disaster mitigation and preparedness as an additional demand on already scarce temporal, human, and financial resources. Second, the decision to install disaster mitigation and preparedness measures is also hindered because it is not practical to prepare for every contingency. Thus, the question remains: How can nonprofits best approach organizational disaster management?

Organizational Survival and Resilience

This chapter places emphasis on nonprofits based on the general presumption that these organizations generally respond more effectively to disasters because they are less encumbered by state bureaucracy or federal policy procedures (Fremont-Smith et al., 2006; Smith, 2006; Green et al., 2007). However, nonprofit organizations need to be able to withstand disaster impact by minimizing losses to life and property if they are to respond to the needs of disaster victims.

Disaster management research suggests at least four basic sets of variables that may influence the survival of businesses or organizations following a natural disaster event. These include organizational characteristics, whether a disaster impact is direct or indirect, the existence of loss containment measures taken predisaster, and whether the organization had any previous disaster experience (Dahlhamer and Tierney, 1998). As noted above, nonprofits in general differ from private businesses or public agencies, especially in the areas of revenue generation, donor influence, and their commitment to the communities they serve. Still, a key question remains as to how the nonprofits that are not primarily engaged in disaster-related missions perceive, prepare for, and respond to disaster events.

While an organization might survive a disaster, its continued viability or resilience, its "ability to recover, spring back, or return to previous circumstances after encountering problems or stresses" or disasters, remains another issue for consideration (Pena et al., 2014). At least six factors are positively associated with organizational resilience: organizational managers' perception of environmental risk, the extent to which the managers seek information about environmental risks, organizational structure, the organization's extent of participation in the community, the degree of compliance with operations continuity plans, and whether the organization has professional accreditation (Somers, 2009).

When considering the context of localized effects of natural disasters on sector resilience, Danes et al. (2009) identified two factors as the most important influences of organizational resilience to natural disasters: the structure of an organization and the extent of its participation in the community. Given the supplementary, substitutionary, and complementary roles nonprofits play in our communities, their resilience also becomes crucial to the response, recovery, and development of local communities (Demiroz and Hu, 2014). Structurally, nonprofits tend to be less financially stable and are at risk of demise following a disaster, which constitutes a serious loss to any community's social capital. Two additional factors also stand out with respect to predictors of organizational survival and resilience: whether an organization has both existing continuity and predisaster plans.

Continuity Planning

Some attention is accorded to the disaster response aspect of continuity planning—also a determinant of organizational resilience—since it transcends different types of risks and "goes beyond emergency response or disaster planning" (Meyer-Emerick and Momen, 2003). Continuity planning is described as a process that focuses on enabling nonprofits to maintain critical operational processes after "natural, anthropogenic, or technological interruptions" have occurred or despite the disaster impact (Meyer-Emerick and Momen, 2003).

Of the nonprofits interviewed in Chikoto et al.'s (2013) study, 30 of the 34 nonprofits in Memphis, Tennessee, reported having contingency or emergency plans in place. Note that of the 34 nonprofits, 11 (32%) were health organizations such as hospitals, home and ambulatory health care services, and psychiatric and substance abuse hospitals, and 7 (20%) were colleges, universities, and some elementary and secondary schools—in short, organizations that are mandated to adopt emergency management plans. Of the 6 (18%) nonprofits engaged in human services fields such as child care, individual, and family services, 5 reported having contingency plans.

While these numbers seem positive, one cannot generalize these to be representative of all nonprofits in a community. Moreover, Meyer-Emerick and Momen (2003) further conceded that continuity planning's "successful development and implementation depends on *preparedness* and innovation and *on management support*," indicating that nonprofit managers still need to prioritize organizational disaster mitigation and preparedness. Moreover, postdisaster recovery and development are influenced by the outcomes of disaster mitigation and preparedness, as well as overall disaster response (Demiroz and Hu, 2014).

Disaster Mitigation and Preparedness

Whereas disaster management research made significant progress in the identification of determinants of disaster mitigation and preparedness adoption at the individual and household levels, research is still lagging behind with respect to determinants of organizational disaster preparedness. For example, within the context of Lindell and Perry's (2004, 2012) Protective Action Decision Model (PADM), *risk perception, disaster experience,* and *resource availability* were found to be positive predictors of the likelihood of households' adoption of hazard adjustments (mitigation and preparedness measures). Meyer-Emerick and Momen (2003) also found a positive correlation between nonprofits' fiscal resources and their ability to adopt continuity planning.

In the area of nonprofit disaster mitigation and preparedness, focusing on yet another determinant of organizational resilience—managerial environmental risk perception—a few comparative studies focused on nonprofit employees' perceptions of organizational risk and preparedness, with research focusing on mid- to high-level leadership (Fowler et al., 2007; Light, 2008). For instance, when comparing nonprofits' top management's perceptions of organizational crisis preparedness to those of their employees, top management held higher perceptions of preparedness than the typical employee (Fowler et al., 2007).

Focusing on five types of crises—natural disasters, major terrorist attacks, secondary terrorist attacks, accidental disasters, and workplace violence—Fowler et al. (2007) found that government agencies expressed higher perceptions of disaster preparedness than nonprofits and businesses. Similarly, in Light's (2008) study, government organizations were perceived to be more crisis-ready than nonprofit organizations (29%) and businesses (20%). While these two studies were informative with regard to management's and employees' perceptions of their organization's level of disaster preparedness, what remains elusive are the types of hazard adjustments or disaster mitigation and preparedness measures nonprofits have in place, especially non-NVOAD members.

Getting closer to this above question, Chikoto et al. (2013) compared Memphis-based nonprofit, private, and public organizations' adoption of a set of disaster mitigation and preparedness strategies (hazard adjustments). The authors concluded that in spite of having constrained resource, nonprofits adopted more hazard adjustments than private businesses. However, as noted above, this finding was based on a sample dominated by health and education nonprofits—the two largest resource powerhouses in the nonprofit sector.

Table 2.1 illustrates the kinds of disaster mitigation and preparedness measures adopted by different types of nonprofits in Memphis, Tennessee. First, while 85% of the nonprofits interviewed reported having

Table 2.1 Percentage and Number (by Key Subsectors) of Nonprofits Adopting Each Hazard Adjustment

	Nonprofit (N = 34)	Education (N = 7)	Health (N = 11)	Human Services (N = 6)
Attended disaster meetings/ training courses outside the organization	84.8%	5	10	6
Held disaster-related workshops/ trainings within the organization	73.5%	6	7	5
Arranged site visits by consultants or experts to better prepare for disasters	38.2%	5	1	2
Provided information to customers/members of the community on issues related to disasters	50.0%	7	2	3
Assessed or evaluated vulnerability to disasters or estimated potential losses from disasters	72.7%	6	6	5
Engaged in nonstructural mitigation measures (e.g., securing computers)	64.7%	5	8	3
Engaged in structural mitigation measures (e.g., strengthening parts of a building)	32.4%	3	2	1
Mentioned a potential disaster in an organizational meeting	76.5%	6	8	5
Discussed in an organizational meeting short-term responses to disasters	73.5%	7	7	4
Discussed in an organizational meeting long-term strategies for recovery from disasters	55.9%	5	5	3

attended a disaster meeting or training course, 36% (N = 10) consisted of nonprofits in the health field. And while a total 38% (N = 13) of the nonprofits reported having arranged site visits by experts or consultants to better prepare for disasters, 38% of these (N = 5) consisted of those in the education field. Of the 74% (N = 25) that reported discussing short-term responses to disasters in an organizational meeting, 28% ($N = 7$) were in the health field, with 28% ($N = 7$) in the education field. And of the 56% ($N = 19$) of nonprofits that reported discussing long-term strategies for recovery from disasters in an organizational meeting, the health and education fields dominated.

While these numbers shed some light on nonprofit disaster mitigation and preparedness strategies, this picture is far from complete. The information contained in this table is not representative of *all* nonprofits in the nation, let alone those nonprofits that are a part of the social safety nets in our communities. Granted, evidence from the literature indicated that some sectors or fields engage in more mitigation and preparedness than others (Sadiq, 2010; Drabek, 1991). Sadiq (2010) found that health and education organizations were more likely to adopt more predisaster measures than the retail and wholesale sectors. This differential may be the same with nonprofits. As such, more representative data are needed to comprehensively address the state of the nonprofit sector's disaster mitigation and preparedness and, in particular, the preparedness level of smaller human service nonprofits.

Furthermore, we also need to analyze the types of mitigation and preparedness measures nonprofits adopt and whether nonprofits are adopting more active or passive measures. It is important to develop an understanding of the nature of the preparedness measures nonprofits are adopting to strengthen their organizational resilience to natural disasters. If there is no preparation, it is equally important to explore the reasons for nonadoption. Such research will provide practitioners and policy makers with information on the barriers and challenges that prevent nonprofits from investing and adopting disaster mitigation and preparedness strategies.

Nonprofits have a place in the nation's emergency management system and are capable of playing essential functions in all four stages of emergency preparedness, mitigation, response, and recovery.

CONCLUSIONS: CHALLENGES AND OPPORTUNITIES

The intensity and frequencies of disasters reveal the challenges government agencies encounter in providing adequate resources to assist disaster-affected communities along the whole spectrum of emergency management—mitigation, preparedness, response, and recovery. Despite experiencing extensive damage themselves, nonprofit organizations are key actors in disaster response, filling in the gap left by government by providing essential complementary and supplementary services to disaster victims, even when disaster response is not their primary mission. Emerging research also demonstrates their importance as key actors in postdisaster recovery and development (Demiroz and Hu, 2014). However, various authors concede that similar to continuity planning, the ability to engage in postdisaster recovery and community development is influenced by both the outcomes of disaster mitigation and preparedness and the disaster response (Demiroz and Hu, 2014; Meyer-Emerick and Momen, 2003).

According to disaster research, the extent of an organization's participation in the community influences its resilience to natural disasters (Danes et al., 2009). Evidently, local nonprofits were and are usually thrust into the front lines following a disaster. In fact, as demonstrated from the disasters noted above, some local nonprofits remained relatively more agile in their response to disaster victims than their government counterparts despite their limited resources. This highlights two strengths that nonprofits and civic associations have over other organizational forms: such organizations are embedded in our local communities and have a stronger pulse on community needs, and they have the capacity to generate resources from multiple sources (government, foundations, and individuals), especially in the aftermath of localized disaster events.

However, in addition to an organization's extent of participation in the community, its structure also influences its resilience to natural disasters (Danes et al., 2009). Hence, nonprofits' capacity to generate resources from multiple sources needs to be appropriately viewed as a double-edged sword. As noted earlier, the nonprofit sector is largely comprised of very small organizations, and these tend to be human service organizations that are responding to local needs in an attempt to address various societal ills. As such, nonprofits' capacity to generate funding from multiple sources needs to be tempered by the fact that individual funding streams are guided by unique motivations, and therefore behave inconsistently (Young, 2006). Additionally, not all nonprofits have the capacity to attract and obtain all funding streams equally due to feasibility issues or mission misalignment (Young, 2006). Furthermore, external funding often comes with various restrictions.

Generally, a sizable number of nonprofits encounter challenges generating sufficient revenue to cover their operations. This, by extension, would most likely hamper any potential disaster mitigation and preparedness efforts. In light of the many response roles nonprofits play following a disaster and the unique position they occupy in our communities, nonprofits are encouraged to find creative ways to generate unrestricted funding for purposes of assessing organizational vulnerabilities to disasters.

Donors, including foundations and government, also need to take tangible steps to emphasize, promote, and invest in nonprofit disaster preparedness to help nonprofits.

However, financial investments in any venture are usually preceded by leadership buy-ins and endorsement. As noted, bolstering organizational resilience is largely influenced by managers' perceptions of environmental risk, to the extent that such managers seek out relevant information, among other factors (Somers, 2009). In exploring the core predictors of organizational disaster preparedness, a random survey of 4,000 opinion leaders in state and local governments, businesses, and nonprofit organizations (468 responded to the survey) conducted by Princeton Survey Research Associates International in 2006 found some organizational characteristics (e.g., leadership and external relationships, etc.) were essential to perceived disaster preparedness or readiness (Light, 2008).

Nonprofits are encouraged to designate responsibility and duties for organizational disaster mitigation and preparedness to an identifiable team of people within the organization. Nonprofit boards of directors, leaders, and managers will need to spearhead and prioritize organizational disaster mitigation and preparedness efforts if a culture of emergency management is to take root among nonprofit organizations. Nonprofits have a unique capacity to attract skilled and unskilled volunteers, from which risk managers and individuals with disaster preparedness experiences can be hired. This is important, as organizations with risk managers tend to adopt more risk-reducing measures than those without such leadership (Sadiq and Graham, 2014).

There are other resources available to help address the disaster mitigation and preparedness needs of nonprofit organizations. One such resource is Collaborating Agencies Responding to Disasters (CARD), a California-based nonprofit with a mission to "prepare local community groups to participate in coordinated response and recovery efforts for vulnerable and under-served populations in Alameda County."* The nonprofit offers preparedness and continuity training and planning, networking opportunities, and other related services. The Department of Homeland Security also provides preparedness grants to "assist states, urban areas, tribal and territorial governments and nonprofit organizations, as well as the private sector to strengthen the nation's ability to prevent, protect against, mitigate, respond to, and recover from terrorist attacks, major disasters, and other emergencies in support of the National Preparedness

* See http://cardcanhelp.org/about-us/mission-and-vision.

Goal and the National Preparedness System."* However, these grants are mostly intended for the "nation's highest risk areas, including urban areas that face the most significant threats."† Nonetheless, nonprofit grantees can utilize these grants to invest in training and updating their disaster planning and procedures.

In view of the importance of external relations as one of the core predictors of organizational crisis preparedness, this chapter calls for increased public–nonprofit partnerships and collaborations—especially among those nonprofits that are least likely to engage in formal collaborations (Guo and Acar, 2005). Partnerships and collaborations are essential not only in disaster response and recovery efforts, but also in ensuring that the nonprofit organizations and voluntary associations embedded in our communities are prepared for disasters, whether or not emergency management is part of their mission. As crucial partners in disaster preparedness, mitigation, response, and recovery, bolstering nonprofits' disaster mitigation and preparedness practices, especially those of human services organizations, cannot be emphasized enough.

REFERENCES

Arlikatti, S., Lindell, M., and Prater, C. (2007). Perceived stakeholder role relationships and adoption of seismic hazard adjustments. *International Journal of Mass Emergencies and Disasters*, 25(3), 218–256.

Anheier, K. H. (2005). *Nonprofit Organizations: Theory, Management, Policy*. New York: Routledge, Taylor & Francis Group.

Auer, J. C., and Lampkin, L. M. (2006). *Open and Operating? An Assessment of Louisiana Nonprofit Health and Human Services after Hurricanes Katrina and Rita*. Washington, DC: Urban Institute and the Louisiana Association of Nonprofit Organizations.

Blackwood, A. S., Roeger, K. L., and Pettijohn, S. L. (2012). *The Nonprofit Sector in Brief: Public Charities, Giving, and Volunteering, 2012*. Washington, DC: Urban Institute.

Boris, E. T., and Steuerle, E. C. (2006). Scope and dimensions of the nonprofit sector. In *The Nonprofit Sector: A Research Handbook*, ed. W. W. Powell and R. Steinberg. New Haven, CT: Yale University Press.

* See http://www.dhs.gov/news/2014/03/18/dhs-announces-grant-guidance-fiscal-year-fy-2014-preparedness-grants.

† Ibid.

Brody, E. (2006). The legal framework for nonprofit organizations. In *The Nonprofit Sector: A Research Handbook*, ed. W. W. Powell and R. Steinberg. New Haven, CT: Yale University Press.

Chikoto, G. L., Sadiq, A. A., and Fordyce, E. (2013). Disaster mitigation and preparedness: Comparison of nonprofit, public, and private organizations. *Nonprofit and Voluntary Sector Quarterly*, 42(2), 389–408.

Comfort, L. K., Waugh, W., and Cigler, B. A. (2012). Emergency management research and practice in public administration: Emergency, evolution, expansion, and future directions. *Public Administration Review*, 72(4), 539–547.

Corporation for National and Community Service. (2014). The State of Volunteering in America, 2014. http://volunteeringinamerica.gov.

CRE. (2013). Impact of Hurricane Sandy on Staten Island nonprofits: A town hall event. Community Resource Exchange, New York, July 30.

Dahlhamer, J. M., and D'Souza, M. J. (1997). Determinants of business disaster preparedness in two U.S. metropolitan areas. *International Journal of Mass Emergencies and Disasters*, 15, 265–281.

Dahlhamer, J. M., and Tierney, K. J. (1998). Rebounding from disruptive events: Business recovery following the Northridge earthquake. *Sociological Spectrum*, 18, 121–141.

Danes, S. M., Lee, J., Amarapurkar, S., Stafford, K., Haynes, G., and Brewton, K. E. (2009). Determinants of family business resilience after a natural disaster by gender of business owner. *Journal of Developmental Entrepreneurship*, 14(4), 333–354.

Demiroz, F., and Hu, Q. (2014). The role of nonprofits and civil society in post-disaster recovery and development. In *Disaster and Development: Examining Global Issues and Cases*, ed. N. Kapucu and K. T. Liou, 317–330. Environmental Hazards Series. Cham, Switzerland: Springer International Publishing.

Drabek, T. E. (1991). Anticipating organizational evacuations: Disaster planning by managers of tourist-oriented private firms. *International Journal of Mass Emergencies and Disasters*, 9, 219–245.

Fagnoni, C. M. (2006). Observations on charities' response: Hurricanes Katrina and Rita. In *After Katrina: Public Expectation and Charities' Response*, ed. E. T. Boris and E. C. Steuerle, 11–16. Emerging Issues in Philanthropy Seminar Series. Washington, DC: Urban Institute.

FEMA. (2008). International lessons to improve mass fatality management during a pandemic influenza event. http://www.drlatulane.org/groups/haiti-recovery/MSM/Mass%20Fatality%20Management%20final.pdf.

FEMA. (2011). National Disaster Recovery Framework: Strengthening disaster recovery for the nation. http://www.fema.gov/pdf/recoveryframework/ndrf.pdf.

Fowler, K. L., Kling, N. D., and Larson, M. D. (2007). Organizational preparedness for coping with a major crisis or disaster. *Business and Society*, 46(1), 88–103.

Fremont-Smith, M., Boris, E. T., and Steuerle, E. C. (2006). Charities' response to disasters: Expectations and realities. In *After Katrina: Public Expectation and Charities' Response*, ed. E. T. Boris and E. C. Steuerle, 1–4. Emerging Issues in Philanthropy Seminar Series. Washington, DC: Urban Institute.

Gajewski, S., Bell, H., Lein, L., and Angel, R. J. (2010). Complexity and instability: The response of nongovernmental organizational organizations to the recovery of Hurricane Katrina survivors in a host community. *Nonprofit and Voluntary Sector Quarterly*, 40(2), 389–403.

Green, J., Kleiner, A., and Montgomery, J. (2007). The texture of local disaster response: Service providers' views following Hurricane Katrina. *Southern Rural Sociology*, 22(2), 28–44.

Guo, C., and Acar, M. (2005). Understanding collaboration among nonprofit organizations: Combining resource dependency, institutional, and network perspectives. *Nonprofit and Voluntary Sector Quarterly*, 34(3), 340–361.

Haimes, Y. Y. (2012). Strategic preparedness for recovery from catastrophic risks to communities and infrastructure systems of systems. *Risk Analysis*, 32(11), 1834–1845.

Hall, P. D. (2004). Historical perspectives on nonprofit organizations in the United States. In *Jossey-Bass Handbook of Nonprofit Leadership and Management*, ed. R. Herman, 3–38. San Francisco: Jossey-Bass.

Hammack, D. C. (2002). Nonprofit organizations in American history. *American Behavioral Scientist*, 45(11), 1638–1674.

Hansmann, H. B. (1980). The role of nonprofit enterprise. *Yale Law Journal*, 89(5), 835–901.

Herman, M. L. (2004). Risk management. In *The Jossey-Bass Handbook of Nonprofit Leadership and Management*, ed. R. D. H. Associates. Vol. 2. Washington, DC: Jossey-Bass Publications.

Herman, M. L., Head, G. L., Jackson, P. M., and Fogarty, T. E. (2003). *Managing Risk in Nonprofit Organizations: A Comprehensive Guide*. Hoboken, NJ: John Wiley & Sons.

Kapucu, N. (2006a). Public-nonprofit partnerships for collective action in dynamic contexts of emergencies. *Public Administration*, 84(1), 205–220.

Kapucu, N. (2006b). Interagency communication networks during emergencies: Boundary spanners in multiagency coordination. *American Review of Public Administration*, 36(2), 207–225.

Kapucu, N., and Wart, M. V. (2006). The evolving role of the public sector in managing catastrophic disasters: Lessons learned. *Administration and Society*, 38(3), 279–308.

Kramer, R. M. (1982). *Voluntary Agencies in the Welfare State*. Berkeley: University of California Press.

Light, P. C. (2008). *Predicting Organizational Crisis Readiness: Perspectives and Practices Toward a Pathway to Preparedness*. New York: New York's Center for Catastrophe Preparedness and Response.

Lindell, M. K., and Perry, R. W. (2000). Household adjustment to earthquake hazard: A review of research. *Environment and Behavior*, 32, 590–630.

Lindell, M. K., and Perry, R. W. (2004). *Communicating Environmental Risk in Multiethnic Communities.* Thousand Oaks, CA: Sage Publications.

Lindell, M. K., and Whitney, D. J. (2000). Correlates of seismic hazard adjustment adoption. *Risk Analysis,* 20 (1), 13–25.

Lindell, M. K., and Perry, R. W. (2012). The protective action decision model: Theoretical modifications and additional evidence. *Risk Analysis,* 32(4), 616–632.

Major, D. (2013). Impact of Hurricane Sandy on Staten Island nonprofits: A town hall event. Staten Island Foundation and Staten Island Not-for-Profit Association. New York: Dara Major Philanthropy Consulting.

Meyer-Emerick, N., and Momen, M. (2003). Continuity planning for nonprofits. *Nonprofit Management and Leadership,* 14(1), 67–77.

Milward, B. H., and Provan, K. G. (2000). Governing the hollow state. *Journal of Public Administration Research and Theory,* 10, 359–379.

Munich Re. (2012). Natural catastrophes worldwide 1980–2012. Princeton, NJ: Geo Risks Research, NatCatSERVICE.

Munich Re. (2014). Loss events worldwide 1980–2013: 10 costliest hurricanes ordered in insured losses. Princeton, NJ: Geo Risks Research, NatCatSERVICE.

NCCS. (2012). Core files. National Center for Charitable Statistics. http://nccs.urban.org/statistics/quickfacts.cfm.

NYNP. (2012). Sandy's impact on nonprofits. New York Nonprofit Press, November 2.

Pena, A. A., Zahran, S., Underwood, A., and Weiler, S. (2014). Effect of natural disasters on local activity. *Growth and Change,* XX(X), 1–21.

Pipa, T. (2006). *Weathering the Storm: The Role of Local Nonprofits in the Hurricane Katrina Relief Effort.* Washington, DC: Aspen Institute.

Rotolo, T., and Berg, J. A. (2010). In times of need: An examination of emergency preparedness and disaster relief service volunteers. *Nonprofit and Voluntary Sector Quarterly,* 40(4), 740–750.

Sadiq, A. A. (2010). Digging through disaster rubble in search of the determinants of organizational mitigation and preparedness. *Risk, Hazards and Crisis in Public Policy,* 1(2), article 2.

Sadiq, A. A., and Graham, J. D. (2014). Organizational risk perception of disasters: Do risk managers matter? Presented at the Risk, Perception, and Response Conference, Harvard Center for Risk Analysis, March 20, 2014.

Salamon, L. M. (1995). *Partners in Public Service: Government-Nonprofit Relations in the Modern Welfare State.* Baltimore, MD: Johns Hopkins University Press.

Salamon, L. M. (2003). The resilient sector: The state of nonprofit America. In *The State of Nonprofit America,* ed. L. M. Salamon. Washington, DC: Brookings Institution Press.

Salamon, L. M., Anheier, H. K., List, R., Toepler, S., and Sokolowski, S. W. (1999). Civil society in comparative perspectives. In *Global Civil Society: Dimensions of the Nonprofit Sector,* ed. L. M. Salamon and H. K. Anheier. Baltimore, MD: Johns Hopkins Center for Civil Society Studies.

Salamon, L. M., Hems, L. C., and Chinnock, K. (2000). *The Nonprofit Sector: For What and for Whom?* Baltimore, MD: Johns Hopkins Center for Civil Society Studies.

Simo, G., and Bies, A. L. (2007). The role of nonprofits in disaster response: An expanded model of cross-sector collaboration. *Public Administration Review*, December (special issue), 125–142.

Smith, D. H. (1997). The rest of the nonprofit sector: Grassroots associations as the dark matter ignored in prevailing "flat earth" maps of the sector. *Nonprofit and Voluntary Sector Quarterly*, 26, 76–80.

Smith, S. R. (2006). Government financing of nonprofit activity. In *Government and Nonprofit Organizations: The Challenges of Civil Society*, ed. E. T. Boris and E. C. Steuerle. Washington, DC: Urban Institute.

Smith, S. R., and Lipsky, M. (1993). *Nonprofits for Hire: The Welfare State in the Age of Contracting*. Cambridge, MA: Harvard University Press.

Somers, S. (2009). Measuring resilience potential: An adaptive strategy for organizational crisis planning. *Journal of Contingencies and Crisis Management*, 17(1), 12–23.

Spillan, J. (2003). An exploratory model for evaluating crisis events and managers' concerns in nonprofit organizations. *Journal of Contingencies and Crisis Management*, 11(4), 160–169.

Steinberg, R. (2006). Economic theories of nonprofit organizations. In *The Nonprofit Sector: A Research Handbook*, ed. W. W. Powell and R. Steinberg. New Haven, CT: Yale University Press.

Syles, R. (2008–2009). Public managers, volunteer organizations, and disasters. *The Public Manager*, Winter, 76–80.

Taylor, I. L., Krane, K. N., Amedee, R. G., and Kahn, M. J. (2006). Rebuilding institutional programs in the aftermath of Hurricane Katrina: The Tulane experience. *Journal of the Medical Sciences*, 332(5), 264–268.

Toepler, S. (2010). Government funding policies. In *Handbook of Research on Nonprofit Economics and Management*, ed. B. A. Seaman and D. R. Young. Northampton, MA: Edward Elgar.

Urban Institute. (2005). The aftermath of Katrina: State of the nonprofit sector in Louisiana. National Center for Charitable Statistics at the Center on Nonprofits and Philanthropy, Office of Public Affairs, Washington, DC.

Vita, C. J. D., and Morley, E. (2007). *Providing Long-Term Services after Major Disasters*. Washington, DC: Urban Institute.

Waugh, W. L. (1999). *Living with Hazards Dealing with Disasters: An Introduction to Emergency Management*. Armonk, NY: M. E. Sharpe.

Waugh, W. L., and Streib, G. (2006). Collaboration and leadership for effective emergency management. *Public Administration Review*, December (special issue), 131–140.

Weisbrod, B. A., and Asch, E. D. (2010). Endowments for a rainy day. *Stanford Social Innovation Review*, Winter.

World Bank. (2010). *Natural Hazards, Unnatural Disasters: The Economics of Effective Prevention*. Washington, DC: World Bank.

Young, D. R. (2000). Alternative models of government-nonprofit sector relations: Theoretical and international perspectives. *Nonprofit and Voluntary Sector Quarterly*, 29(1), 149–172.

Young, D. R. (2003). How nonprofit organizations manage risk. In *Paid and Unpaid Labour in the Social Economy*, ed. S. Destefanis and M. Musella. Heidelberg, Germany: Physica-Verlag.

Young, D. R. (2006a). Complementary, supplementary or adversarial: A theoretical and historical examination of government-nonprofit relations in the U.S. In *Government and Nonprofit Organizations: The Challenges of Civil Society*, ed. E. T. Boris and E. C. Steuerle. 2nd ed. Washington, DC: Urban Institute.

Young, D. R. (2006b). Why study nonprofit finance? In *Financing Nonprofits: Putting Theory into Practice*, ed. D. R. Young. Lanham, MD: Altamira Press.

Zhang, Y., Lindell, M., and Prater, C. (2009). Vulnerability of community businesses to environmental disasters. *Disasters*, 33(1), 38–57.

3

Country Mouse, City Mouse

Exploring the Differences in Rural and Urban Economic Recovery Postdisaster

Davia Cox Downey

Contents

INTRODUCTION

> What I'm hearing which is sort of scary is that they all want to stay in Texas. Everybody is so overwhelmed by the hospitality. And so many of the people in the arena here, you know, were underprivileged anyway so this (chuckle)—this is working very well for them.
>
> —Former First Lady Barbara Bush, on the hurricane evacuees at the Astrodome in Houston, September 5, 2005*

In September 2005, the deadliest and most destructive hurricanes hit the Atlantic seaboard, making landfall in the state of Louisiana. Hurricane Katrina caused an estimated $81 billion worth of damage to the Gulf Coast, and Hurricane Rita, which followed a few weeks later, caused an additional $4 billion to $5 billion worth of damage (Bratton and Haynie, 1999; Bratton et al., 2006; Burby, 2006; Department of Homeland Security, 2006; Saulny, 2006). Nearly 2,000 lives were lost during the storms, and an additional 250,000 residents were displaced after the disaster, some permanently. In the wake of these natural disasters, disaster preparedness, mitigation, and postdisaster local revitalization took center stage as federal, state, and local government entities took stock and created plans to rebuild the Gulf region.

Scholarship on Hurricane Katrina is particularly concerned with the city of New Orleans and its response to managing the aftermath of the storms. New Orleans is certainly worth this type of rigorous analysis; however, this chapter expands the focus to the entirety of the affected Gulf Coast area and is unique because it draws conclusions that are applicable for rural and state policy makers in terms of economic disaster recovery. Another subject of investigation by social scientists is the operational failures of the Federal Emergency Management Agency (FEMA) and the overall failures of the emergency management systems in place before the storms hit (Eikenberry et al., 2007; Gerber, 2007; Petak, 1985; Schneider, 1995, 2005). Much of the scholarship to date, particularly on Hurricane

* Barbara Bush, "Marketplace NPR Interview," September 5, 2005, http://www.marketplace.org/shows/2005/09/05/PM200509051.html.

Katrina, has focused on failures of emergency disaster planning, namely, FEMA's lack of response (Eikenberry et al., 2007; Gerber, 2007; Petak, 1985; Schneider, 2005), the race and class divide after the storm (Elliott and Pais, 2006; Lavelle and Feagin, 2006; Stivers, 2007; Tynes et al., 2006), and difficulties that local and state governments face when attempting to rebuild after a natural disaster (Eckdish-Knack, 2006; Lewis, 2005; Liu, 2006; Olshansky, 2008). But public administration and policy scholars have largely been silent on issues of race and poverty in the aftermath of Hurricane Katrina. What has also been missing in this discussion is an examination of what less populated areas within the state of Louisiana did to recover economically. In particular, this chapter looks specifically at the effect of poverty on economic recovery in the entire state of Louisiana post-Katrina.

While some scholars have examined the effect of race on recovery in New Orleans (Baade et al., 2007; Comfort, 2006; Elliott and Pais, 2006), discussion of the storm's impact on other areas in the Gulf is not widely in evidence. Instead of focusing on the economic development of the city of New Orleans, this research compares the economic behavior of the city to that of the rural communities (i.e., counties) also affected by the storms.

This chapter also considers whether there is any evidence of systematic variations in the way communities recovered after the hurricane season of 2005. In particular, questions about poverty and its effect on economic development post-disaster are asked to ascertain whether there is a separate effect that can be attributed to this variable. Finally, this chapter assesses the effect of disasters on rural economies, a critical area of study often overlooked by scholars of emergency management and disaster policy.

LITERATURE REVIEW AND THEORY

Counties as Engines of Economic Growth

Many communities in the Gulf region suffer from high levels of poverty. Following the storm, media reports highlighted many issues that faced the area both before and after the 2005 hurricane season. The most striking to many outside observers was the intersection of race and class in the wake of a natural disaster (Bates, 2006; Casserly, 2006; Congleton, 2006). In the United States, race and class are closely intertwined, and public policy has, at times, promoted institutionalized racism (Better, 2008; Leon, 1979). In terms of economic health, cities with higher proportions of white, middle-class residents generally enjoy more stable housing and job markets and have less political complicated policy issues to debate. In places where there is a severe lack of an educated workforce coupled with high proportions of the working poor or minority residents, the literature repeatedly uncovers systematic problems in housing and economic development (Vergara, 1999a, 1999b; Olshansky, 2006; Newman, 2008; Rogers, 2008).

County-level governance, particularly in rural or sparsely populated areas, suffers from the same issues as those who control the political power and for what purpose they use the power to affect public policy. Counties typically coordinate public services, levy taxes to create area wealth, and work to draw new business to the area. Traditionally, counties perform state-mandated duties, which include property assessment, record keeping (e.g., property and vital statistics), road maintenance, election administration, and poor relief (National Association of Counties, 2008). Today, counties are moving into other areas, undertaking programs related to economic development, employment/training, planning and zoning, and water quality.

Competition, Cooperation, and Rural Governance

The earliest research on local governance and policy discusses the issues of pluralism and elitism as they affect the policy-making process (Abney and Lauth, 1985; Cooper et al., 2005; Davis et al., 1997; Truman, 1951; Mills, 1956). The theory and literature that follow here discuss some critical issues that affect emergency response in disasters. As this research is concerned specifically with how race and class may interact with the initial emergency management policy-making process and subsequent recovery after a natural disaster has passed, complementary theories of cooperation and competition, bias in policy making, and the difficulties in planning for natural disasters in general will be reviewed.

The argument presented in this chapter is that heterogeneous areas (those with high concentrations of minorities and less educated and poor residents) have more difficulty determining what issues should be on the policy agenda, as opposed to their more homogenous counterparts, resulting in uneven emergency planning for disasters (Nemeth, 1986; Nemeth et al., 1974; Park and Vargas-Ramos, 2002). This heterogeneity also plays a significant role in how areas recover economically since intergroup conflict before a disaster may be exacerbated after disaster. Thus, it is assumed that most areas with high levels of poverty will have difficulties not only in determining how emergency planning and execution should occur, but also in recovering economically poststorm.

The earliest research in politics debates the importance of power (Cooper et al., 2005; Dahl, 1961; Hunter, 1953; Kettl, 1993; Moe, 2005; Ordeshook and Schwartz, 1987). The question of who gets what and when remains central in most political research. The connection to public policy lies in who maintains political power within cities, counties, states, and nations. The majority has the power to dictate the direction of policy, and if the minority is not cohesive or large enough to make a significant dent in the desires of the majority, public policy will be slanted toward majority desires (Nelson, 1979; Oliver, 2000; Pelissero, 2003).

Early pluralist scholars noted that in the United States, ethnic integration is unavoidable in a pluralistic democracy (Dahl, 1961). As minority groups gained their independence, public policies changed to meet the demands of these newly enfranchised groups. Pluralists determined that in a representative democracy, in order to have policy preferences realized, ethnic groups should compete through electoral contests and economic development in order to be assimilated into the mainstream of urban political life. Additionally, those who occupy the lower classes within the United States are typically disorganized when it comes to political participation. Myriad factors, such as crowded policy agendas, lack of access to policy makers, and conversely, a lack of outreach effort directed at understanding many important political issues, result in policies that may not necessarily benefit these groups. In the case of county governance, where voting coalitions and neighborhood cohesion are particularly feeble, one can only surmise that this issue would be more problematic.

When looking at class variables in the local context, we find that local government institutions also affect the ability of citizens to access the political system. Banfield and Wilson's *City Politics* (1963) was one of the first forays to explore the effects of municipal reforms. Their political ethos theory pitted private-regarding immigrant working class values, which favored sustaining traditional political machines, against public-regarding white Anglo Protestant middle-class values, which supported municipal reform. Thus, lower-class workers had very different policy preferences than white-collar counterparts, but institutional reforms made it difficult for poorer residents to participate in the political process. This theory also left little role for African Americans in the municipal reform movement since, at the time, African Americans could not vote. Since the adoption of municipal reforms like the city manager–council form of government, short ballots for local elections, nonpartisan ballots, and professionalized management of public services, scholars examining black voter turnout have often found low rates of participation by blacks and other minority groups (Oliver, 2000; Verba, 1993; Wolfinger, 1965; Wood, 2002). Additionally, those in the lower economic classes also had dismal voting participation rates.

Cohen and Dawson (1993), in their exploration of the effects of neighborhood poverty, found that the devastation of the urban core, as first described by Banfield and Wilson (1963), propagated a decline in black voting, particularly in areas that are well below the poverty level. This decline affects blacks in particular, because decades after the massive influx of freed slaves during emancipation, the city core continues to be populated by African Americans and poor whites who do not have the economic resources to move from the city center (Banfield and Wilson, 1963). This situation decreases the effectiveness of institutions that promote political behavior in these areas (community groups, party offices, or churches that support political candidates), thus reducing the political efficacy of residents.

J. Eric Oliver and Tali Mendelberg (2000; Oliver, 1999, 2000, 2001) studied in depth the effects of socioeconomic status, city size, and segregation on political participation. In considering the social composition of American localities, Oliver convincingly showed the impacts on the dynamics of democratic government. His findings indicate that individual political behavior varies systematically across economic contexts. He does not find a positive relationship between affluence and participation; instead, he finds that political participation is lower in cities with higher median household incomes and in those that are more economically homogeneous. Therefore, while homogeneous municipalities participate less, they also seem to have fewer things to debate. What this theory means, contextually, is that in rural America, it is assumed that the economic and political value systems of residents are largely the same, regardless of one's racial or ethnic background, so poverty is assumed to trump all other types of social circumstance. This research tries to unpack these assumptions and test this colloquialism using economic data over a sustained period of time.

Governmental Response Postdisaster

The issue, in regard to whether coalitional or competition-based politics is at work, speaks to the heart of this query. When a natural disaster occurs, the operational capacity of government is threatened. As groups begin to reassert themselves after such an event, there is a question of whether new groups are able to permeate the political system for policy change. Schneider (1995) theorized that the key to successful governmental response after a natural disaster depends on how much postdisaster human behavior corresponds to expectations. Thus, if response patterns do not correspond to preplanned structures, disconnects (both politically and socially) occur. If we know that minority groups (in this case, this attribute could be either a racial or class construct), regardless of what type of theoretical perspective adopted, have a difficult time affecting the political and policy processes, then we can expect their economic outcomes after a natural disaster to be different from those of their more educated, richer counterparts as per Oliver's theory.

This effect should be prevalent in counties with high levels of poor residents. The best way to examine this issue is to make a direct comparison between counties with high levels of the poor and those with low levels of the poor. Conversely, if we expect the premise of group theory to hold, places with high levels of poverty will suffer from the same problems of political participation and influence on economic recovery postdisaster. The area of emergency planning and disaster relief policy would also be affected by a lack of consideration of these groups' needs. Hurricane Katrina is a prominent example of this disconnect: once local emergency response capacity was overloaded, the failure of disaster response continued up to the federal level. Thus, the question remains, after a natural disaster, how does policy respond and whom does it respond to?

As stated above, the economic health of an area is integral to its survival. Without jobs, residents cannot buy or rent homes, buy goods within a jurisdiction, or pay taxes. Thus, tracking the types and number of jobs created postdisaster is one indicator of policy response. The question of those economically disadvantaged to affect their own circumstance postdisaster can be tracked by looking at changes in an area's class makeup in terms of employment after a disaster.

Theory: Rural versus Urban Economic Development

There is some literature that deals with the effects of minority and poverty population on the economic outputs of a locale and the effects of diversity, which are critical to the theory developed in this analysis, because emergency management policy is connected to returning economic normalcy in an area once a disaster has passed. The theoretical perspective presented here has two parts. The first concerns the racial composition of a locale, which connects to the power structure theories of local politics; the second uses the frame of poverty to test similar assumptions that in either case act as large impediments to economic recovery postdisaster and cannot be overcome with a large influx of funding. Consistent with the group theory of policy making, it is hypothesized that poverty acts as a deterrent to cooperation, resulting in postrecovery measures favoring one group over another. Thus, this study seeks to determine the ratio of poverty-stricken residents within a locale to the number of jobs created postdisaster.

Alesina and La Ferrara (2005) surveyed the literature on the economics of diversity and found that the production of public goods in ethnically and racially fragmented societies, regardless of the size of the location, be it a city, county, or village, is lower than that in places where racial and ethnic homogeneity is present. Additional work by Alesina and LaFerrara (2000, 2002) determined that trust and social capital are critical to ameliorating issues of public goods production. Looking at economic outputs and demographic fragmentation, the authors found that private investment is lower in places that have high fragmentation.

Paul Collier (2000) found that diversity is tied to lower overall economic growth in places where political rights are limited. Given the South's history of suppressing the political participation of minorities in various ways, and the current positions of both Louisiana and Mississippi at the lower ends of economic investment compared to other states in the nation,* the conclusions Collier presented are worth mentioning here. Finally, Ferraro and Cummings (2007) found the economic behavior in societies is directly related to differences in ethnicity, race, and religion. Research into the effects of income stratification has yielded similar results (Tomaskovic-Devey and Roscigno, 1996; Moller et al., 2009). Thus, higher fragmentation on racial, ethnic, and poverty lines yields lower economic investment and outputs at the state, city, and country levels (Ottaviano and Peri, 2005).

There are several factors that contribute to economic growth in rural areas: levels of taxation, public spending, wage levels, unionization levels, proximity to higher education, access to highways, airports, and other transportation, per capita or family income, and population size and density, among others (Aldrich and Kusmin, 1997; Kusmin, 1994). One study found that the percentage of African Americans living in rural areas

* A Wallethub report in 2014 showed that in terms of taxpayer return on investment (ROI), Louisiana ranked 49th in providing infrastructure, public services, educational resources, and economic stability out of all 50 states (http://wallethub.com/edu/state-taxpayer-roi-report/3283/).

contributed to slow growth, as the study empirically tested the social and economic variables listed above on economic outputs in rural areas throughout the United States (Aldrich and Kusmin, 1997). Additionally, stagnant growth in per capita income in rural areas is consistently found to contribute to overall sluggish economies in rural areas (Porter, 2004, 2005; Rozelle, 1996). Much like the theories of group participation and racial competition inform the debate over policy development in local government, these same theories of group participation, organization, and mobilization of bias in decision-making patterns also shed light on how local economic decisions are made in rural and urban areas. This research will explore the effectiveness of economic recovery in two areas with very different racial and poverty compositions predisaster, focusing on the impacts of state and federal investment in the wake of such an exogenous shock. It will also explore the dynamics of economic recovery in rural and exurban communities located far from the central core cities during Hurricane Katrina utilizing the same metrics of racial and class composition.

HYPOTHESES

Based on the literature discussed and the case comparison at hand, the following hypotheses are tested:

- H1: Higher percentages of African Americans decrease the amount of economic recovery postdisaster.
- H2: Higher percentages of residents under the poverty line decrease the amount of economic recovery postdisaster.
- H3: Higher levels of outside federal and state funding produce a more robust postdisaster economic recovery, regardless of racial and economic disparities.
- H4: Rural places will have decreased economic recovery in comparison to urban or suburban places.

DATA AND METHODS

Case Descriptions

For this analysis, all 37 parishes designated by FEMA as eligible for receiving public or individual assistance were selected, and a full listing of the parishes included in the analysis is listed in the appendix. This analysis compares economic recovery data for New Orleans and the state of Louisiana from 2004 to 2010. For New Orleans, the seven-parish metropolitan statistical area (MSA) is used to make direct comparisons with the rest of Louisiana. The parishes included in this group are Jefferson, Orleans, Plaquemines, St. Barnard, St. Charles, St. Tammany, and St. John the Baptist.* The population of these seven parishes accounts for over 1.3 million residents.

New Orleans' poverty rate in 2000 was 28% compared to 12% for the entire nation (U.S. Census, 2008). This rate dropped to 23.4% from 2005 to 2008, while the average rose to 18.4% in the United States. In 2010, the overall poverty rate for the city was 24.1% (U.S. Census Bureau, 2011). The black poverty rate was more than three times the white poverty rate (35% vs. 11%), and 43% of blacks lived in high-poverty neighborhoods in 2010.

The dissimilarity index, a measure of racial segregation, ranks New Orleans high among large-city minority populations. Blacks and whites live in different worlds, marked by geographical separation as well as other measures of socioeconomic status. At the time Katrina hit, New Orleans was one of the poorest and most segregated cities in America. Residential segregation was so prevalent that in order for the city to have an equal distribution of blacks and whites in every neighborhood within the city, 69% of black residents would have had to relocate. Additionally, poverty played a large role in the stratification of New Orleans, with the city experiencing a 12% unemployment rate in 2004, which at the time of Katrina was double that of the nation (Holzer and Lerman, 2006).

In Louisiana, the racial composition is very different from that in New Orleans. Overall, the state is whiter (approximately 63% compared to 55% in New Orleans). In terms of per capita income, Louisiana residents earned approximately $22,500 on average during the years 2005–2009 compared to the average per capita income of $27,000 in the United States for the same time period. Compared to residents of New Orleans, whose average income was $25,000 during the same time period, rural residents earned approximately 14% less than their urban counterparts. The state's poverty rate in 2000 was 17.2% and rose to 21.6% in the year 2010 (U.S. Census, 2013). The National Assessment of Educational Progress (2013), which measures several education outcomes, consistently ranked Louisiana below average in terms of literacy; basic proficiency in reading, writing,

* A table of all cases, including population and other socioeconomic variables, can be found in Table 3.1.

science, and math; and overall educational attainment (i.e., high school and college graduation rates).

Variables

Studies of aggregate-level economic impacts have used many methodologies to determine the short- and long-term effects of natural disasters on the economy. Some studies have used indicators such as the employment or unemployment rate, the number of small businesses created postdisaster, personal income changes, tax receipts, and gross regional product (Skidmore and Toya, 2002, 2013; Tierney, 1997, 2012; Tierney and Oliver-Smith, 2012; Xiao, 2011). This work largely shows that in the aggregate, areas are fairly resilient to natural disasters, but they do little to ascertain the more localized effects of natural disasters.

The main dependent variable—the net number of new jobs—is defined by the census as employees in newly created jobs who were not employed by the same employer in the previous quarter. This information is divided by the total number of jobs in each place minus the total number of job separations (i.e., firings, leaves of absence, or downsized positions) for each year and each location in the data set to create a rate of net new hires.

$$\text{Net new hire rate} = \frac{\text{New jobs added} - \text{Job separations}}{\text{Total jobs}}$$

This indicator was chosen as the measure of economic recovery for several reasons. First, it acts as an index variable because it simultaneously tracks the new job activity within an area, job separations, jobs created, and total jobs in each local economy for each quarter throughout the time period under examination as a dynamic percentage, meaning that it changes from quarter to quarter with a range of –0.50 to 0.50, indicating a net job shift of half a percentage in either direction. This information is provided by the Quarterly Workforce Indicators survey as part of the Longitudinal Employee Household Dynamic data set for the Census; it is frequently used to determine overall economic performance in terms of the strength (or weakness) of the American labor market.

There are some limitations to this measure. First, the way the census collects these data is dependent on county and state agencies reporting information in a timely manner, and in some cases, certain counties or parishes did not submit their information, and therefore no economic data could be entered for certain periods. Second, the measure does not track individuals identified by the census as being singularly affected by the storms. The census does have a subset of individuals surveyed post-Katrina, but these data weren't collected on a quarterly basis after the storm, nor did they include all residents in the state; thus, they cannot be included here.

Rural Economic Data

The Local Economic and Household Dynamics (LEHD) and Local Employment Dynamics (LED) data sets, as well as the Bureau of Labor Statistics (BLS), collect data only periodically from sparsely populated places, making an analysis of rural recovery very difficult. While not a comprehensive measure of economic recovery, new hires provide a viable relative measure on which municipalities can be compared because all counties have to report this information at least once a year to the LEHD and the BLS, the measure comes directly from businesses located within each county (i.e., it is not collected from secondary sources), and numbers are verified by each state's bureau of labor before being sent to the BLS.

For each geographic entity included in the data set, the new jobs variable represents the estimated number of workers who started a new job (Abowd et al., 2011, pp. 4–6). Specifically, new hires represents the total number of workers that, though they worked for an employer in the specified quarter, were not employed by that same employer in any quarter of the previous year. The total jobs component of the dependent variable reflects the beginning of the quarter estimate of the total number of jobs in the economy on the first day of the reference quarter (Abowd et al., 2011). Thus, a worker is counted in the total jobs figure if he or she has positive earnings in both $t-1$ and t. Finally, the job separation variable is calculated as the estimated number of workers who had a job for at least one previous full quarter and then the job ended (Abowd et al., 2006, pp. 75–77). The number is an estimate due to the fact that actual separation data are not available until the following quarter, when reports from each state are received by the census. Thus, a worker is defined as separated if he or she had positive earnings in t or $t-1$, but no earnings in $t+1$; this identifies the separation in the current quarter distinct from the previous one.

New hires as a variable are used in other research measuring recovery from natural disasters. For example, Brookings and the Greater New Orleans Community Data Center (GNODC) have tracked various recovery indicators since the disaster, including, but not limited to, new firm development, average wages, and increases and decreases in various employment sectors (Liu et al., 2011). Although Brookings followed additional indicators, such as gross metro product per job created, nonfarm jobs by sector, and average annual wages over a 6-year period, these were not viable for use in this study due to the fact that the Brookings Institute has done little comparative work on rural parishes. Additionally, most data on the counties and parishes located outside the New Orleans MSA are collected sporadically by the census, and therefore researchers have shied away from conducting this type of work.

Independent Variables

The main independent variables for this study are the amount of turnover experienced in each location during the time period under examination (turnover), the yearly percentage of black residents in a county, the percentage of residents living under the poverty level in each county for the years under study, the percentage of residents with a high school diploma, and the federal and state investment postdisaster. The percentage of minority residents is measured as the percentage of African American, non-Latino residents in each location for each year in the data set. Included as control variables are the average black earnings for all jobs in the area under examination, average earnings of residents with less education than a high school diploma, and high school graduate earnings, as well as the overall percentage of new hires for the same categories (i.e., the number of black earnings in period 1, the high school graduate earnings in the period, and so on). The regression models were estimated by sorting the counties in terms of their urban–rural continuum codes as recorded in the 2003 census by the U.S. Department of Agriculture (USDA). These are coded as ordinal variables ranging from 0, representing urban counties with 1 million or more residents in the metropolitan area, to 3, representing nonmetropolitan counties with less than 25,000 residents adjacent or nonadjacent to metropolitan counties. This variable was added to establish whether the amount of "rural-ness" throughout the affected state had a significant difference on the amount of job recovery postdisaster. The percentage of residents living in poverty is taken directly from the Small Area Income and Poverty Estimates data set. Finally, a regression model was run substituting the average new hire earnings for three populations: African American, those residents with less than a high school diploma, and those with a high school diploma. The idea was to ascertain the differential effects of minority and less educated job seekers over the same time period.*

Federal disaster funds are the total amount of funds approved for each county during the period 2006–2007. This information came from the office of Research and Special Projects housed in the Louisiana Recovery Authority (Office of Community Development, 2008) and through a Freedom of Information request to FEMA. The dollar amounts include individual and household assistance (IA), which also include FEMA's Individuals and Households Program (IHP) and public assistance (PA) funds dedicated to recovery in each county. IHP assists homeowners and renters with reconstruction after a declared disaster (FEMA, 2008). Funds help applicants pay for temporary housing, repair, replacement, and disaster-related medical services, burial costs, and moving expenses. FEMA's public assistance (PA) grant program directs funding to states, local governments, tribal areas, and certain nonprofit organizations to assist in helping them respond during and after a declared disaster. For this study, the percentage of each state's total funds received by each parish/county was calculated. Another control variable measuring catastrophic housing damage collected by FEMA in collaboration with the Small Business Administration and the U.S. Department of Housing and Urban Development (HUD) in 2006 for each parish was also included.

The Louisiana Road Home program was designed as a separate state-level effort to provide payment to homeowners affected by Hurricane Katrina. It offered up to $150,000 in compensation to homeowners for losses sustained after the hurricanes. It also provided reparation to owners of rental properties in the form of loans and grants for rebuilding. The amount of funds contributed under the Road Home program was recorded for each county. Again, the percentage of total funding received for each parish was used. For the years affected (2005–2009), the variable is coded 1, and the previous years are coded 0. A year dummy variable was added to ascertain the magnitude of difference of the race or poverty effect on new jobs for each year since the storm had passed. The appendix lists all the variables used in the research as well as their source.

* It should be noted here that the U.S. Census collects data on either race and ethnicity attributes or educational attainment for all new hires in each period, but does *not* collect the data together. This makes it impossible to ascertain whether a new hire is both African American and has less than a high school diploma as one "hire." For this reason, all new higher earning data are separated out by either race or educational attainment for each period.

METHODS AND ANALYSIS

The first part of the analysis compares counties based on their rural continuum score. The definition of rural that guides this analysis is taken from their classification on the rural–urban continuum.

This analysis employs descriptive summary statistics, reports the Pearson's correlation coefficients to test the linear relationships between race, poverty, and funding within the state, and incorporates ordinary least squares (OLS) regression using robust standard errors for each county over the 8-year period. OLS regression typically assumes that the residuals are independent. In the case of this analysis, the data collected on each county or parish during the time period may not be independent, thus violating the independent principle of regression that states that errors associated with one observation are not correlated with the errors of any other observation. In this case, it is likely that job creation within one parish will tend to follow a predetermined path, or be more like one another than the kinds of jobs created in a parish located right next door. A robust standard error indicates that the observations are clustered by some unit, in this case county/parish, and that the observations may be correlated *within* units, but would be independent *between* county and parish. The use of robust standard errors, or the Huber–White estimation, relaxes the assumption that errors are independent and identically distributed. Essentially, the use of the clustered function in an OLS regression model is to make adjustments in the estimates that take into account some of the flaws in the data (Davidson and MacKinnon, 1993; Wooldridge, 2010).

Prior to delving into the analyses, a comment about the role of race and class as separate variables in the analysis should be made. Traditionally, researchers have utilized race as a proxy for class as African Americans are traditionally overrepresented among socioeconomically disadvantaged groups (Banfield and Wilson, 1963; McIlwain and Caliendo, 2011; Gilens, 1999, 2012). Class, typically measured by poverty level in the census, acts as a severe impediment to housing and jobs and also affects residential mobility. However, because poverty levels are measured only in the decennial census for many areas, and some of the more rural counties included in this analysis have no reliable data for poverty levels throughout the period under examination, high school education levels are used as a proxy.

Race and class as variables, however, are separate phenomena and are appropriate to analyze separately in the following analyses for three reasons. First, the majority of the poor in the United States are whites. Second, using race without class in analyses or vice versa encourages the illusion that class confounds racial differences. Race, in the United States, is an ascriptive characteristic and acts as an antecedent to class in various settings (i.e., job placement, college admissions, housing, etc.) (Rank, 2004). Finally, in sociological research as well as in the health professions, race often acts as a "caste" within American politics and policy making, limiting access to goods and service; class is largely an invisible quality that acts as an impediment to services or access but is not a visible indicator of discrimination (Cox, 2000; Kawachi et al., 2005).

FINDINGS

New Orleans and the State of Louisiana

By comparing the MSA* to the state and the counties that are most urbanized within the affected area, we see that the New Orleans area had a higher proportion of minorities, suffered greater damage during and after the storm, and received more individual and public assistance. Table 3.1 shows the wide variation in funding received between the New Orleans MSA and the rest of the state, while Table 3.2 presents the summary statistics of minority and poverty percentage, percentage major damage as reported by HUD, and percentage of federal and state funding received by each parish.

Per capita funding received by residents outside the New Orleans MSA (i.e., mostly urban parishes) was less than half that received by residents inside the city. Additionally, within the New Orleans metropolitan statistical area there are slightly higher percentages of minorities and slightly lower percentages of poor residents (see Table 3.2). Housing damage was the greatest in New Orleans; the New Orleans MSA also received the most outside investment from federal and state-level programs.

Table 3.1 Per Capita Expenditures by State[a]

	FEMA Funds	State Funds	Per Capita
Suburban and rural parishes	$8,139B	$896M	$4,711.97
New Orleans metropolitan area	$6,546B	$6,271B	$11,553.53

[a] Through FY 2007

* The New Orleans MSA includes the following parishes: Jefferson, Orleans, Plaquemines, St. Bernard, St. Charles, St. John the Baptist, and St. Tammany.

Table 3.2 Descriptive Statistics of Affected Parishes

	Minority Ratio (%)	Class Ratio (%)	Housing Damage (%)	FEMA IA Funds (out of 100%)	FEMA PA Funds (out of 100%)	Road Home Funds (out of 100%)	Rural–Urban Continuum Code
Ascension	22	12	16	0	0	0	0
East Baton Rouge	46	19	11	1	1	0	0
East Feliciana	46	22	17	0	0	0	0
Iberville	51	22	21	0	0	0	0
Jefferson	30	15	53	22	8	16	0
Lafayette	27	16	29	0	1	0	0
Livingston	6	13	22	0	0	0	0
Orleans	68	25	72	54	42	47	0
Plaquemines	29	16	80	2	8	2	0
Point Coupee	38	22	11	0	0	0	0
St. Bernard	12	17	81	7	24	12	0
St. Charles	29	13	50	1	1	0	0
St. Helena	53	22	45	0	0	0	0
St. John the Baptist	49	17	46	0	0	0	0
St. Tammany	13	11	71	8	11	0	0
Vermillion	17	19	39	0	0	1	0
West Baton Rouge	37	17	14	0	0	0	0
West Feliciana	51	21	9	0	0	0	0
Iberia	35	21	26	0	0	1	0

(Continued)

Table 3.2 Descriptive Statistics of Affected Parishes (Continued)

	Minority Ratio (%)	Class Ratio (%)	Housing Damage (%)	FEMA IA Funds (out of 100%)	FEMA PA Funds (out of 100%)	Road Home Funds (out of 100%)	Rural–Urban Continuum Code
Acadia	20	22	25	0	0	0	0
Calcasieu	26	17	64	0	0	5	1
Cameron	5	14	90	0	0	1	1
Lafourche	17	16	29	0	1	0	1
Tangipahoa	30	23	44	1	1	1	1
Terrebonne	25	18	32	0	0	1	1
Allen	27	22	45	0	0	0	2
Evangeline	30	27	13	0	0	0	2
Sabine	25	19	16	0	0	0	2
St. James	50	17	35	0	0	0	2
Assumption	33	20	25	0	0	0	2
Beauregard	15	26	51	0	0	0	2
Jefferson Davis	19	19	50	0	0	0	2
St. Landry	43	27	16	0	0	0	2
St. Martin	33	20	15	0	0	0	1
St. Mary	36	22	24	0	0	9	2
Vernon	20	17	24	0	0	0	2
Washington	32	25	68	1	4	0	2

Moving to the regression analysis results, the analysis proceeds with two models (see Table 3.3). The first model estimates the effects of the funding and damage variables, while the second model represents the estimates without these variables. This procedure was employed for two reasons. First, utilizing the clustered regression model allows us to group observations by parish or year, but not both. Therefore, once these variables were found to be nonsignificant in a number of iterations, it was determined that they should be dropped. Second, and more importantly, the variables measuring damage and funding received by parishes from FEMA or the Road Home program were recorded in 2006 due to an artifact in the data received from the government agencies from which the data were requested. Both reports showed the total numbers of dollars given, not the year or month in which these dollars were distributed. So in effect, these variables are artificially time bound due to data restrictions.

We can summarize the following trends in economic recovery post-Katrina throughout the affected parishes. First, in terms of economic recovery, the amount of job turnover in each parish, regardless of urbanity, is the most significant factor in determining robust recovery throughout the examined time period. Next, for suburban communities, both the percentage of African American residents and the number of residents employed with less than a high school education also dictated the strength of economic recovery. In urban areas, the average earnings of the black population had a significant and negative drag on economic recovery, and in the suburban areas, again education; in this case, higher numbers of those employed with just a high school diploma negatively impacted economic recovery. Interestingly, in rural parishes, none of the variables of interest, save job turnover, had significant effects in either model. The overall fit of the models for urban parishes was moderate, and although the R^2 values for the suburban parishes are high in the first model, we should interpret this with caution, as this particular category had very few observations for both clustered regression models.

DISCUSSION AND CONCLUSIONS

Hypothesis 1—whether high proportions of African Americans within a parish had a negative effect on the net new jobs created—was confirmed in the urban parishes under observation. In suburban parishes, however, economic recovery was also statistically affected by high percentages of African American residents, but only when the time-bound variables

Table 3.3 Two Models of Urban, Suburban, and Rural Economic Recovery

	Model 1			Model 2		
	Urban	**Suburban**	**Rural**	**Urban**	**Suburban**	**Rural**
Percentage of job turnover	–1.086***	–0.943	–0.501	–0.783***	–0.927**	–0.719***
	(–0.1239)	(–0.3105)	(–0.6494)	(–0.0504)	(–0.0867)	(–0.0889)
Percentage of high school graduates	–0.0000169	–0.0053	–0.00159	0.000328	0.0000474	0.000722
	(–0.0009)	(–0.0097)	(–0.0017)	(–0.0004)	(–0.0013)	(–0.0005)
Percentage of African American non-Latinos	0.444	–0.732	0.0228	0.0997	–0.216**	–0.00134
	(–0.3643)	(–0.4659)	(–0.0949)	(–0.0918)	(–0.0174)	(–0.0413)
Percentage of residents under poverty level	–0.0133	–0.0287	–0.00369	–0.00233	0.000631	0.00171
	(–0.0103)	(–0.0259)	(–0.0074)	(–0.0022)	(–0.0028)	(–0.002)
Percentage of FEMA individual assistance	1.518	0	–28.01			
	(–1.5403)	(.)	(–24.4229)			
Percentage of FEMA public assistance	–3.563	0	0			
	(–3.6414)	(.)	(.)			
Percentage of Road Home assistance	–1.528	0	0.389			
	(–1.5374)	(.)	(–0.3664)			
Percentage of catastrophic housing damage	0.0129	0	0.0118			
	(–0.0106)	(.)	(–0.0059)			

(Continued)

Table 3.3 Two Models of Urban, Suburban, and Rural Economic Recovery (Continued)

	Model 1			Model 2		
	Urban	**Suburban**	**Rural**	**Urban**	**Suburban**	**Rural**
Black new hire average earnings	−0.000191*	0.0000192	0.000209	−0.00000999	0.0000626	−0.000071
	(−0.0001)	0	(−0.0002)	0	0	0
Less than high school diploma average earnings	0.0000813	−0.00163*	−0.000301	0.00000983	0.0000865	0.0000171
	(−0.0001)	(−0.0004)	(−0.0003)	0	(−0.0001)	(−0.0001)
High school graduate average earnings	0.000087	0.00157*	0.0000911	−0.0000156	−0.000103	0.0000937
	(−0.0001)	(−0.0005)	(−0.0002)	0	(−0.0001)	(−0.0001)
Constant	0.16	1.443	0.198	0.0381	0.106	−0.0743
	(−0.0766)	(−1.5465)	(−0.2099)	(−0.029)	(−0.0772)	(−0.0548)
Number of clustered observations by rural continuum code	68	16	28	352	80	140
R-squared	0.414	0.774	0.228	0.231	0.253	0.179
Adjusted *R*-squared	0.299	0.576	−0.227	0.216	0.181	0.135

Note: Standard errors are below the coefficients in parentheses. All models estimated using clustered standard regression.
p values: $^{*}p < .05$; $^{**}p < .01$; $^{***}p < .001$

measuring emergency management funds distributed and catastrophic housing damage were removed from the model. In terms of residents living under the poverty line, the analyses did not show a significant effect on net new job creation in any urban–rural continuum classification. Finally, regarding the fourth hypothesis, economic recovery when controlling for disaster assistance was not significant in any case.

Based on the findings just discussed, several conclusions are drawn. In New Orleans, race, rather than class, does appear to be related to new job creation postdisaster. In other words, larger African American populations appear to hinder disaster recovery, particularly in very urban places. Inside and outside the New Orleans metropolitan area, federal and state assistance dollars also had no statistical effect.

Overall, high poverty does not seem to affect job recovery postdisaster. However, it should be noted that once the sample was split, significances in terms of education level and average earnings based on race and education level reveal themselves. In all, it seemed that the disaster assistance sent throughout the state of Louisiana and to residents (and public agencies who qualified for public assistance) in New Orleans had no effect on the creation of new jobs. These dollars helped in other ways, but the impact of funding was not captured by the job recovery data tested here. Taken together, these findings suggest that racial diversity challenges economic recovery after a disaster more than poverty, but only in areas that have relatively less racial diversity overall (e.g., rural areas and less segregated metropolitan areas). In short, while postdisaster media coverage dramatically captured the plight of African American residents trapped in New Orleans, minorities in more isolated areas suffered greater damage in the long term.

There should be some discussion on why public assistance dollars appear to have had a larger impact on the economic fortunes of these places than state disaster programs did, as revealed in the state of Mississippi results. Public assistance dollars were funds directed at local governments, nongovernmental organizations (NGOs), quasi-public entities, school districts, and other units of government to repair infrastructure damaged after the hurricanes. Additionally, existing contracts, staff capacity, and expertise and organizational infrastructure may have aided these public organizations in promoting recovery. Additionally, these organizations and local government units receiving PA funds could have been directly responsible for job creation. Why this effect was not seen in Louisiana is a puzzle that should be explored in future research. There could have been additional policy-related factors or other unobserved variables that were at play in the economic fortunes of the state of Louisiana than are presented here. Additionally, future analyses that look at the quality of the job market prior to and following a natural disaster would be of interest to researchers interested in economic development postcrisis.

What can be learned from the analysis here? First, high concentrations of African Americans have depressing effects on job creation postdisaster in certain circumstances. Second, federal and state funding programs have very little effect on helping areas to recover quickly postdisaster unless they are targeted or include funding to organizations and governments as well as individuals. The mixed results of funding also imply that not enough money was spent on disaster recovery or it was not spent effectively. Or, they indicate that funding in general had a limited effect on job creation due to the fact that the path by which money was sent to local governments to spend on job creation was not linear. Local government units may not necessarily use money received postdisaster to create new positions; instead, they use it in other ways (i.e., to replace infrastructure or equipment, or to pay salaries of returning workers) not directly connected to new job creation. The private sector saw the most new job creation postdisaster, but in rural parishes, the longevity and quality of these jobs are difficult to ascertain due to data constraints.

Policy Implications

What does this mean for policy? In considering the use of federal funds, policy makers should consider the use of targeted funding for rural communities and focus on providing programs for the development of economies in these sparsely populated areas. Further, it suggests that policy makers and planners need to explicitly consider concentrated relief programs for minorities and those with less access to quality education in more remote communities. Stabilization of earnings and the quality of the job market both matter; therefore, once recovery efforts that are directed at reconstruction are complete, policy makers and local leaders should ask themselves what residents are doing to keep themselves gainfully employed postdisaster. Policy makers involved in the mitigation stage of emergency management planning should focus on individual job sectors and their vulnerabilities to recovery when disaster strikes. Additional focus on those industries that employ the most minorities in rural and urban areas should also be inventoried to better serve these sectors. Preparing emergency managers and policy makers at the local, state, and federal levels by making them aware of the critical need to focus effort in rural economic development is necessary to prepare better for future disasters.

ACKNOWLEDGMENTS

Data analysis support was from Prof. Laura A. Reese at Michigan State University, Prof. Emeritus Gary Sands at Wayne State University, Prof. Shane Singh at University of Georgia, and Prof. William Myers at University of Tampa. Additional editorial support was from Renata Horvath and Gregory King of Grand Valley State University.

REFERENCES

Abney, G., and T. P. Lauth. 1985. Interest group influence in city policy-making: The views of administrators. *Western Political Quarterly* 38(1):148–161.

Abowd, J., B. Stephens, and L. Vilhuber. 2006. *Detailed QWI Definitions*. Washington, DC: U.S. Census.

Abowd, J., B. Stephens, and L. Vilhuber. 2011. *LED: Quarterly Workforce Indicators 101*. Washington, DC: U.S. Census.

Aldrich, L., and L. Kusmin. 1997. *Rural Economic Development: What Makes Rural Communities Grow?* Agriculture Information Bulletin No. (A1B-737), 8 pp. Distributed by ERIC Clearinghouse.

Alesina, A., and E. La Ferrara. 2000. Participation in heterogeneous communities. *Quarterly Journal of Economics* 115(3):847–904.

Alesina, A., and E. La Ferrara. 2002. Who trusts others? *Journal of Public Economics* 85(2):207–234.

Alesina, A., and E. La Ferrara. 2005. Ethnic diversity and economic performance. *Journal of Economic Literature* 43(3):762–800.

Banfield, E. C., and J. Q. Wilson. 1963. *City Politics*. Cambridge, MA: Harvard University Press.

Bates, L. K. 2006. Post-Katrina housing: Problems, policies, and prospects for African-Americans in New Orleans. *The Black Scholar* 36(4):13–31.

Better, S. J. 2008. *Institutional Racism: A Primer on Theory and Strategies for Social Change*. 2nd ed. Lanham, MD: Rowman and Littlefield Publishers.

Bratton, K. A., and K. L. Haynie. 1999. Agenda setting and legislative success in state legislatures: the effect of gender and race. *Journal of Politics* 61(3):658–679.

Bratton, K.A., K. L. Haynie, and B. Reingold. 2007. Agenda setting and African American women in state legislatures. *Women, Politics, and Public Policy* 28 (Summer/Fall):71–96.

Burby, R. G. 2006. Hurricane Katrina and the paradoxes of government disaster policy: Bringing about wise governmental decisions for hazardous areas. *Annals of the American Academy of Political and Social Science* 604:171–191.

Casserly, M. 2006. Double jeopardy. In C. Hartman and G. D. Squires (eds.), *There Is No Such Thing as a Natural Disaster: Race, Class, and Hurricane Katrina*. New York: Routledge.

Cohen, C. J., and M. Dawson. 1993. Neighborhood poverty and American politics. *American Political Science Review* 87(2):286–302.

Collier, P. 2000. Ethnicity, politics and economic performance. *Economics and Politics* 12(3).

Congleton, R. D. 2006. The story of Katrina: New Orleans and the political economy of catastrophe. *Public Choice* 127(1–2): 5.

Cooper, C. A., A. J. Nownes, and S. Roberts. 2005. Perceptions of power: Interest groups in local politics. *State and Local Government Review* 37(3):206–216.

Dahl, R. A. 1961. *Who Governs? Democracy and Power in an American City.* Yale Studies in Political Science 4. New Haven, CT: Yale University Press.

Davidson, R., and J. G. MacKinnon. 1993. *Estimation and Inference in Econometrics.* New York: Oxford University Press.

Davis, J. H., W. T. Au, L. Hulbert, X. Chen, and P. Zarnoth. 1997. Effects of group size and procedural influence on consensual judgments of quantity: The example of damage awards and mock civil juries. *Journal of Personality and Social Psychology* 73:703.

Department of Homeland Security. 2006. *Civil Defense and Homeland Security: A Short History of National Preparedness Efforts.* Washington, DC: Department of Homeland Security.

Eckdish-Knack, R. 2006, January. Let the rebuilding begin. *Planning.*

Eikenberry, A. M., V. Arroyave, and T. Cooper. 2007. Administrative failures and the international NGO response to Hurricane Katrina. *Public Administration Review* 67(s1):160–170.

Elliott, J. R., and J. Pais. 2006. Race, class, and Hurricane Katrina: Social differences in human responses to disaster. *Social Science Research* 35(2):295–321.

Federal Emergency Management Agency (FEMA). 2008. *Disaster Assistance,* ed. Department of Homeland Security. Washington, DC: Federal Emergency Management Agency.

Ferraro, P. J., and R. G. Cummings. 2007. Cultural diversity, discrimination, and economic outcomes: An experimental analysis. *Economic Inquiry* 45 (2):217–232.

Gerber, B. J. 2007. Disaster management in the United States: Examining key political and policy challenges. *Policy Studies Journal* 35(2):227–238.

Gilens, M. 1999. *Why Americans Hate Welfare: Race, Media, and the Politics of Antipoverty Policy.* Chicago: University of Chicago Press.

Gliens, M. 2012. *Affluence and Influence: Economic Inequality and Political Power in America.* Princeton: Princeton University Press.

Holzer, H., and R. I. Lerman. 2006. *Employment Issues and Challenges in Post-Katrina New Orleans.* Washington, DC: Urban Institute.

Hunter, F. 1953. *Community Power Structure: A Study of Decision Makers.* Chapel Hill: University of North Carolina Press.

Kettl, D. F. 1993. *Sharing Power: Public Governance and Private Markets.* Washington, DC: Brookings Institution.

Kusmin, L. D. 1994. Factors associated with the growth of local and regional economies: A review of selected empirical literature. In *Staff Report*. Washington, DC: U.S. Department of Agriculture, Economic Research Service, Agriculture and Rural Economy Division.

Lavelle, K., and J. Feagin. 2006. Hurricane Katrina: The race and class debate. *Monthly Review* 58:52–67.

Leon, D. J. 1979. Institutional Racism and the Educational Opportunity Program: A Study of Organizational Change and Strategies for Reform. Presented at the Annual Meeting of the American Sociological Association, Boston, MA, 14 pp.

Lewis, S. 2005, December. Next steps in New Orleans. *Planning.*

Liu, A. 2006. Building a better New Orleans: A review of and plan for progress one year after Hurricane Katrina special analysis in metropolitan policy. Washington, DC: The Brookings Institute, 47 pp.

Liu, A. 2011. *Resilience and Opportunity: Lessons from the U.S. Gulf Coast after Katrina and Rita*. Washington, DC: Brookings Institution.

McIlwain, C., and S. M. Caliendo. 2011. *Race Appeal: How Candidates Invoke Race in U.S. Political Campaigns*. Philadelphia: Temple University Press.

Mills, C. W. 1956. *The Power Elite*. New York: Oxford University Press.

Moe, T. M. 2005. Power and political institutions. *Perspectives on Politics* 3(2):215–233.

Moller, S., A. S. Alderson, and F. Nielsen. 2009. Changing patterns of income inequality in U.S. counties, 1970–2000. *American Journal of Sociology* 114(4):1037–1101.

National Assessment of Educational Progress. January 1, 2013. Annual Report Card. Accessed April 29, 2015. http://nces.ed.gov/nationsreportcard.

National Association of Counties. 2008. *Overview of County Government*. Washington, DC: National Association of Counties.

Nelson, D. C. 1979. Ethnicity and socioeconomic status as sources of participation: The case for ethnic political culture. *American Political Science Review* 73:1024.

Nemeth, C. J. 1986. Differential contributions of majority and minority influence. *Psychological Review* 93:23.

Nemeth, C. J., M. Swedlund, and B. Kanki. 1974. Patterning of the minority's responses and their influence on the majority. *European Journal of Social Psychology* 4:53.

Office of Community Development, Louisiana Recovery Authority. 2008. Baton Rouge. http://www.doa.louisiana.gov/cdbg/cdbg.htm.

Oliver, J. E. 1999. The effects of metropolitan economic segregation on local civic participation. *American Journal of Political Science* 43(1):186–212.

Oliver, J. E. 2000. City size and civic involvement in metropolitan America. *American Political Science Review* 94(2):361–373.

Oliver, J. E. 2001. *Democracy in Suburbia*. Princeton, NJ: Princeton University Press.

Oliver, J. E., and T. Mendelberg. 2000. Reconsidering the environmental determinants of racial attitudes. *American Journal of Political Science* 44:574.

Olshansky, R. B. 2006. Longer view: Planning after Hurricane Katrina. *Journal of the American Planning Association* 72(2):147–154.

Olshansky, R. B. et al. 2008. Planning for the rebuilding of New Orleans. *Journal of the American Planning Association* 74(3):273.

Ordeshook, P. C., and T. Schwartz. 1987. Agendas and the control of political outcomes. *American Political Science Review* 81(1):179–200.

Ottaviano, G. I. P., and G. Peri. 2005. Cities and cultures. *Journal of Urban Economics* 58(2):304–337.

Park, D. K., and C. V. Vargas-Ramos. 2002. Paradigms of minority political participation in the United States. *Research in Micropolitics* 6:253.

Pelissero, J. P. 2003. *Cities, Politics, and Policy: A Comparative Analysis.* Washington, DC: CQ Press.

Petak, W. J. 1985. Emergency management: A challenge for public administration. *Public Administration Review* 45(special suppl.):3–7.

Porter, K. M. 2005. Going broke the hard way: The economics of rural failure. *Wisconsin Law Review* 25(4):969.

Porter, M. E. 2004. Location, competition, and economic development: Local clusters in a global economy. *Economic Development Quarterly* 14(1): 15–34.

Rogers, D. 1971. *The Management of Big Cities: Interest Groups and Social Change Strategies.* Beverly Hills, CA: Sage Publications.

Rozelle, S. 1996. Stagnation without equity: Patterns of growth and inequality in China's rural economy. *China Journal* 35:63–92.

Saulny, S. 2006, July 30th. Despite a city's hopes, an uneven re-population. *The New York Times.* http://www.nytimes.com/2006/07/30/us/30return.html?pagewanted=all&_r=0.

Schneider, S. K. 1995. *Flirting with Disaster: Public Management in Crisis Situations, Bureaucracies, Public Administration, and Public Policy.* Armonk, NY: M. E. Sharpe.

Schneider, S. K. 2005. Administrative breakdowns in the governmental response to Hurricane Katrina. *Public Administration Review* 65(5):515–516.

Skidmore, M., and H. Toya. 2002. Do natural disasters promote long-run growth? Economic Inquiry 40(4): 664–687.

Skidmore, M., and H. Toya. 2013. Natural disaster impacts and fiscal decentralization. *Land Economics* 89(1): 101–117.

Stivers, C. 2007. So poor and so black: Hurricane Katrina, public administration, and the issue of race. *Public Administration Review* 67(s1):48–56.

Subramanian, S. V. et al. 2005. Covariation in the socioeconomic determinants of self rated health and happiness: A multivariate multilevel analysis of individuals and communities in the USA. *Journal of Epidemiology and Community Health* 59(8): 664–669.

Tierney, K. J. 1997. Business impacts of the Northridge earthquake. *Journal of Contingencies and Crisis Management* 5(2): 87–97.

Tierney, K. 2012. Disaster governance: Social, political, and economic dimensions. *Annual Review of Environment and Resources* 37: 341–363.

Tierney, K.J., and A. Oliver-Smith. 2012. Social dimensions of disaster recovery. *International Journal of Mass Emergencies and Disasters* 30(2): 123–146.

Tomaskovic-Devey, D., and V. J. Roscigno. 1996. Racial economic subordination and white gain in the U.S. South. *American Sociological Review* 61(4):565–589.

Truman, D. B. 1951. *The Governmental Process: Political Interests and Public Opinion.* 1st ed. New York: Knopf.

Tynes, B., C. Hunter, and H. A. Neville. 2006. Bush doesn't care about black people: Race, class, and attributions of responsibility in the aftermath of Hurricane Katrina. *The Black Scholar* 36(4):32–42.

U.S. Census. 2008. Small area income and poverty estimates. Washington, DC: U.S. Census.

U.S. Census. 2013. *Historical Poverty Tables—People,* ed. Current Population Survey. Washington, DC: U.S. Census.

U.S. Census Bureau. 2011. Louisiana Quick Facts from the U.S. Census Bureau. Accessed April 29, 2015. http://quickfacts.census.gov/qfd/states/22000.html.

Verba, S. et al. 1993. Race, ethnicity, and political resources: Participation in the United States. *British Journal of Political Science* 23: 453.

Vergara, C. J. 1999a. *The New American Ghetto.* New Brunswick, NJ: Rutgers University Press.

Vergara, C. J. 1999b. *American Ruins.* New York: Monacelli Press.

Wolfinger, R. E. 1965. The development and persistence of ethnic voting. *American Political Science Review* 59: 896.

Wood, C. 2002. Voter turnout in city elections. *Urban Affairs Review* 38(2): 209–231.

Wooldridge, J.M. 2009. *Introductory Econometrics: A Modern Approach.* Mason, OH: South Western, Cengage Learning.

Xiao, Y. 2011. Local economic impacts of natural disasters. *Journal of Regional Science* 51(4): 804–820.

Section II

Transportation Considerations

4

Improving City Resistance in War

Planning Major Transportation Terminals Based on Passive Defense Considerations

Amir Shakibamanesh and Mahshid Ghorbanian

Contents

INTRODUCTION

Nowadays, the emergence of long-range weapons has extended the fighting from the borders to the depth of territories, that is, urban and rural areas, and the traditional boundary between the front and nonfront areas has faded away. In such a situation, in case of enemy attack, citizens behind the front will face as many risks as the soldiers on the battlefield will. In the early hours of the enemy attack, cities will face a huge crisis.

Experiences obtained from past wars, especially the 8-year Iran–Iraq War, the first Persian Gulf War, the North Atlantic Treaty Organization (NATO) bombing of Yugoslavia in 1999, the U.S. and Britain invasion of Iraq in 2003, the Second Lebanon War in 2006, the U.S. war in Iraq, and other conflicts around the world, have confirmed this view that the attacker tries to break the will of the people and the political, economical, and military power of the target country via undertaking strategies to destroy important centers and places, especially those in cities. This is usually done by bombardment (Movahedi Nia, 2008).

During the 8-year Iraq–Iran War, among almost 496 cities, 6 major Iranian strategic cities were razed to the ground. In addition, 17 other cities were damaged—between 15% and 85%—as a result of the enemy's artillery attacks, air raids, and missiles. During this war, 61 Iranian cities were damaged by military attacks. These attacks were not confined to urban areas: 1,138 villages in the country were completely destroyed, and 2,344 villages incurred heavy damage throughout the war (Abbaszadeh Fard, 1999).

Observing the requirements of passive defense in urban planning and urban design plays a great role in reducing cities' vulnerabilities, reducing financial and criminal damage, increasing the threshold of civilians' resistance in acute attack situations, and facilitating urban crisis management. As railways, subways, and airports are among the most important transportation terminals in a city, they are discussed in detail in this chapter.

PASSIVE DEFENSE: A FORGOTTEN CONCEPT IN URBAN PLANNING AND DESIGN

Before addressing safe urban design based on passive defense, *passive defense* should first be defined in order to specify important points in available definitions. According to various sources and published literature by the U.S. Department of Defense, *passive defense* is "a set of non-military measures taken to reduce vulnerability and to minimize the potential damages caused by invading forces" (Chairman of the Joint Chiefs of Staff,

2006; AIPD, 2008). Different measures taken by veterans and military forces in passive defense include early warning; security operations; dispersion policy; protection of important persons and the general public; medical assistance, especially to counter the deadly effects of nuclear, biological, and chemical weapons of mass destruction (WMDs); recovery of forces; disseminating news and information; staff training; and other techniques, tactics, and processes effective in reducing damage caused by attacks.

In Iranian military and strategic texts, *passive defense* is "a series of nonviolent actions with the purpose of increasing resistance against the enemy's attacks on living areas, improving maintenance of essential activities in cities and villages, improving national resistance, and facilitating crisis management against the enemy's military threats and actions" (Movahedi Nia, 2008; Asghryan Jeddi, 1996; Ziari, 2001). The use of passive defense measures reduces casualties and the levels of vulnerability and damage to critical military and civilian buildings, facilities, and equipment, and protects arteries of the country against enemy attacks. It could be helpful in reducing the risks of unnatural incidents.

In some sources, the term *civil defense* is used as an equivalent for *passive defense* and is defined as follows:

> Civil defense is an effort to protect the citizens of a state (generally noncombatants) from military attacks via using the principles of emergency operations, prevention, mitigation, preparation, response, emergency evacuation, and recovery. Programs of this kind were initially discussed at least as early as the 1920s, but only became widespread in the USA after the threat of nuclear weapons was realized. (Baker, 1978)

According to this definition, civil defense guarantees the safety of the civilian population in wartime (Cristy, 1974). Thus, civil defense aims at the following objectives systematically:

- Minimizing the effects of military attacks on civilian populations
- Dealing promptly with emergencies resulting from such attacks
- Retrieving and restoring damaged facilities and services as a result of such attacks (Kummer and Kummer, 1973)

According to this definition, the term *civil defense* is equivalent to the term *passive defense*. However, it should also be mentioned that since the end of the Cold War, some countries shifted their focus of civil defense from military crises to encompass all hazards and crises in general. Thus, a gap has emerged between passive defense and civil defense in current academic sources (Shakibamanesh and Fesharaki, 2011).

THE MAIN FACTORS AFFECTING ATTACKS ON CITIES

Cities are big and stationary targets in physical terms; they are easily targeted from a huge distance. Other factors motivating the choice of cities as military targets are as follows:

- Cities include aggregation of humans and major political, administrative, and military decision-making centers.
- Attacks on cities put politicians and officials under pressure.
- Attacks on cities disrupt social order and create dissatisfaction in the community.
- Attacks on cities break the unity of people and force them to leave the cities.
- Attacks on cities concern soldiers about the situation behind the front.
- Attacks on cities break the resistance of support management forces stationed in cities.
- Having welfare and livelihood facilities and services, cities play a very effective supportive role in guiding and managing wars.
- A significant proportion of physical and cultural investments is done in cities. Therefore, demolition or elimination of access to these assets will strengthen financial incentives in the invading forces.
- Because cities are considered models of stability, taking hold of them plays an important strategic role and is considered an index to prove the military position and operational authority of the invading forces.
- Cities work as transportation links and centers of regional integration for their rural areas. The government status depends on thr resistance or fall of these cities (Shakibamanesh and Fesharaki, 2011; APA, 2006).

Besides the above-mentioned factors, the role of psychological factors and the effects of attack on people's morale in cities are important. The social life in cities and the alignment and proximity of construction make cities an indispensable target for destroying people's morale. For example, during World War II, the cities of London, Paris, and Berlin were heavily bombed to weaken their social and military morale. The most striking example of this type of targeting is the atomic bombing of the cities Hiroshima and Nagasaki by the United States that caused Japan to surrender. In addition, in recent wars, examples of threats and bombings

Figure 4.1 Cities destroyed during World War II. (From Glaser, L.B., Recalling the Origins of *Slaughterhouse-Five, Ezra,* 2011, p. 18, http://ezramagazine.cornell.edu/SPRING11/People.html.)

against cities are conspicuous: the Kosovo War, first Persian Gulf War, Iraq–Iran War, and U.S. War on Terror (Figure 4.1).

DESIGNING AND PLANNING URBAN USES BASED ON MAJOR TRANSPORTATION TERMINALS

Designing and planning optimum urban land uses plays an important role in reducing vulnerability against possible enemy attacks and invasions. The debate on urban use issues seeks answers to a specific question: How should different urban uses be located in or out of an area so that they satisfy the needs of user groups in peace and war situations and their shortcomings and incorrect distribution do not negatively affect these needs, lead to increased losses, or make compensation for consequent injuries and damage difficult?

It should be pointed out that many urban uses should always be located based on the classification and hierarchy of urban spatial structures. For example, a clinic is suitable for an area of residential neighborhoods that are 650 to 750 meters away from the clinic and have populations of between 2,000 to 4,000 people. However, a hospital as a higher-scale land use is suitable for an urban region that is 1 to 1.5 kilometers away from the hospital and has a population of about 10,000 to 14,000 people (Ziari, 2002).

Similarly, when distribution of business uses necessary for satisfaction of public daily needs (such as retail stores, supermarkets, bakeries, etc.) in the city is not scale specific, satisfying the easiest and everyday living needs of residents in large cities is problematic and difficult in acute situations. Different physical–spatial levels of Iranian cities—residential buildings, housing complexes, neighborhood units, areas and districts—have different processes of change in population, spatial structure, uses, and services that should be tailored to these categories in terms of position and deployment. Some urban land uses are considered the main targets of enemy attacks and invasions; therefore, they are very important because loss or destruction of one of them (depending on their type—vital,[1] critical,[2] and important[3]) will have extensive harmful effects and consequences at national, regional, and municipal levels. In fact, the destruction and damage of such uses can lead to huge economic losses and paralyze the industrial and manufacturing structure of the related city, region, or country. In most cases, incorrect locating and establishing of these land uses will increase the level of casualties and damage in their immediate physical environment. Some of these important land uses are railways, subways, and airports.

Railways

Railways are among the main elements of cities and are considered one of the key infrastructures that draw attention in wartime. Competent managers and policy makers responsible for developing and constructing such centers should take into account appropriate measures, preferably before the construction phase, and consider passive defense criteria to reduce the vulnerability of these facilities against possible threats. In this way, by improving urban infrastructures' resistance threshold, they can save them and prevent loss of assets economically.

Railways are important means of transporting goods and passengers long distances; therefore, they are invaluable as a transportation system. Unlike airplanes, the rail system is designed to transport greater volumes of people and goods in a longer time frame with lower costs.

The railway is also considered an important and valuable urban infrastructure that plays a significant role transporting troops, ammunition, weapons, and equipment, as well as food and relief supplies in times of crisis and war. Since World War II, when the railway was used to transport troops and equipment, it has always been among the first and most vulnerable targets of enemy attacks (Shakibamanesh and Fesharaki, 2011). Railways are an integral part of today's cities, and in spite of the fact that they are inflexible in terms of location and mobility, we will provide some solutions in terms of passive defense in order to reduce their vulnerabilities and the negative effects exerted on the surrounding area.

Planning and Design Considerations

- Railway stations should not be placed in cities' central areas. This is discussed from two viewpoints; first, the linearity of a railroad network in many cities acts as a major obstacle to urban development and impedes designers and planners in the process of directing urban development. On the other hand, in terms of passive defense, locating railway stations within the urban context leads to serious damage to surrounding environments in case of attack, and therefore increases casualties and damage. Locating railroads and train stations outside cities and towns, taking into account various developmental concerns, and ensuring that urban development will not be directed to these areas in the future have positive effects in terms of urbanization and decrease the amount of damage to the surrounding environment.
- Since the railroads located within cities are easily distinguishable and recognizable in satellite images due to their linearity and visibility, if the network enters an urban environment, it should be made invisible through specific measures, at least partially. This is done by taking advantage of natural conditions and camouflaging and redirecting the rail network as possible. Dense vegetation planted along the train routes can be used for this purpose.
- In the cases of railroads that enter cities (e.g., Mashhad railroad) and those that pass across urban margins (e.g., Tehran city railroad that passes southern areas of the city), there is always a need to create a safe buffer around railway stations and the surrounding areas (Figures 4.2 and 4.3). This buffer can include a dense green belt of plantings within which ravines are created to prevent or reduce the extension of explosions and fires. These buffers not only have a role as urban green spaces, but also provide safe population accommodation areas in crisis and war conditions (Figure 4.4).

In many cities, especially in small towns through which railroads pass, a major road is occasionally created parallel to the railroad in front of the station. However, because attacks to the railway and station can lead to the destruction of this main road, it is recommended that multiple roads parallel to the main one are designed so that in times of crisis they can be used as alternative routes for travel. The overall city structure should not become linear around the railroad (Figure 4.5).

In many cases, trains stopped at railway stations and their related rail networks are subject to damage and destruction by attack. Thus, in

Figure 4.2 Railway of Mashhad in the central part of the city. (From Shakibamanesh, A., and H. Fesharaki, S.J., *Urban Design from Passive Defense Vision* [in Persian], Boostane Hamid Press, Tehran, Iran, 2011.)

addition to very heavy financial damage, the railway network system is paralyzed in these conditions. The use of underground tunnels—like what is seen in subway systems—and their connection to the main railroad network is useful. These tunnels can be a safe place for stopping trains and provide a means of reestablishing network connectivity and letting urban railroads continue to operate during crises.

Today, in major Iranian cities (including Tehran, Mashhad, Tabriz, Isfahan, and Shiraz), subway projects are in progress or in the final stages (IARTE, 2014). If we can connect the subway and intercity railroad networks, this will provide several benefits in terms of passive defense. First, relocating and evacuating urban populations and transferring them to nearby cities in crisis situations will become possible. Second, trains from the subway system enter the railroad system and vice versa; therefore, destruction of surface trains will not amount to the destruction of the suburban railway network.

Figure 4.3 Tehran railway located in the southern part of the city. (From Shakibamanesh, A., and H. Fesharaki, S.J., *Urban Design from Passive Defense Vision* [in Persian], Boostane Hamid Press, Tehran, Iran, 2011.)

Railways are important high-traffic public spaces; taking measures to create safe havens at railway stations can, to a remarkable extent, reduce human casualties.

Subways

Today subways play a significant role in transporting people within cities, and thereby contribute considerably to urban transport networks, especially in large cities. In this part of the chapter we investigate subway systems in terms of passive defense and discuss this particular means of transport and its utilization in crisis situations. Two of the technical and scientific characteristics of subway construction are that it is located at a level under cities' surface land and that subways operate via underground interconnected networks. Construction of underground networks naturally requires digging the ground and creating certain infrastructure

Figure 4.4 Safety buffer around the Mashhad Railway Station. (From Shakibamanesh, A., and H. Fesharaki, S.J., *Urban Design from Passive Defense Vision* [in Persian], Boostane Hamid Press, Tehran, Iran, 2011.)

facilities and equipment tailored to this transportation network. This requires huge amounts of investment in terms of money and time because creating such a network requires designing, planning, and implementing extensive underground network projects, as well as establishing the best connection possible with overground stations and structures. From planners', urban designers', and traffic engineers' points of view, this connection includes finding the favorable locations and congested areas (where people have major access problems due to lack of a subway system, large distances from the major centers, or failure of public transport networks to cover these areas) and connecting them through subway stations to the underground railway network. As mentioned above, the first strategy to follow in locating subway stations is to find the population concentration areas in cities. These areas are mostly urban, regional, or district centers

Figure 4.5 It is necessary to design more than one main road parallel to the railway, Maraghe City. (From Shakibamanesh, A., and H. Fesharaki, S.J., *Urban Design from Passive Defense Vision* [in Persian], Boostane Hamid Press, Tehran, Iran, 2011.)

that are used by other surrounding areas because of specialized uses present within their immediate environment.

Connections between subway networks should be established in a way that covers maximum areas across the city in order to provide service to different parts of the city and complement other networks of public transportation. Hence, the distribution of main stations and their connections to each other displays the key trajectory of secondary stations. On the other hand, according to subway network design standards, subway stations should be placed in certain specified intervals (about 450 to 1,000 meters) to better transport people from subnodes to main areas attracting urban population (Shakibamanesh, 2009). Figure 4.6 shows the distribution and coverage of subway stations in Tehran.

As evident in the images of Tehran subway stations, the main centers from east to west and from north to south are connected together via networks that are now operating. Subway networks can, in many

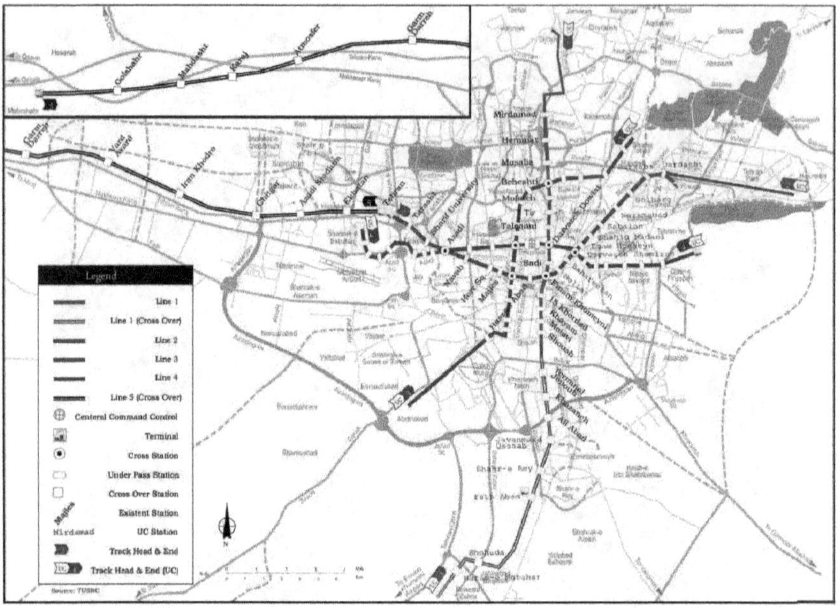

Figure 4.6 Tehran subway network. (From Tehran Municipality, Tehran Metro Map, 2014, http://metro.tehran.ir.)

cases, satisfy the critical needs related to passive defense. This topic is discussed below.

Planning and Design Considerations

The main geometric structure of subways consists of two basic components: nodes (centers) and connecting lines (rail networks). Nodes are subway stations across the city that gather people and guide them into the subway network and distribute the population in urban areas. Because of their layout and location, these nodes are used by many people. In other words, it is possible to gather a lot of people from surrounding areas in a short time frame. So, from this perspective, subway stations are the most desirable places to gather population and keep them safe underground. This feature is used to protect the defenseless urban population present in surrounding areas.

The depth of subway stations in their connection points with overground urban areas makes it possible to use these stations as urban shelters in times of crisis and enemy attacks. Indeed, being placed underground is a perfect criterion for protecting populations from explosions and damage. By including enough depth when designing subway stations, in addition to relevant structural considerations, we can also build nuclear shelters therein. One of the famous historic examples of using this consideration occurred in London during World War II, which is known as Londoners during the Blitz.[4]

The third category includes equipment and installations that must be present in subway stations. The subway is a modern urban space, and because of the complex systems used therein, it must be able to meet the technical and technological requirements of a small underground city. Therefore, its installations and equipment, such as emergency power generators for air conditioning and firefighting systems, play an important role in ensuring individuals' safety in crisis situations. In addition to the above facilities, subways as major urban shelters should have arrangements to store emergency water, fuel, and power. The important thing about subway equipment and facilities is having minimum dependence on the overground facilities, especially in crisis situations. Maximum safety criteria should be considered when constructing and installing equipment and furniture in subway interior spaces.

It should be mentioned that subway stations can accommodate and provide shelter for a large number of people because of their huge underground spaces. Dividing panels and lightweight, inexpensive partitions should be available in subway stations to be installed in critical conditions to divide space in order to meet the psychological needs of individuals based on their social circumstances and culture (Figure 4.7).

Locating and deploying subway stations and their surrounding uses should be done in a way in which they are placed near urban critical uses, such as chain stores, large supermarkets, and so forth, that provide services to people present in subway stations. Special spaces in subways should also be allocated for this purpose. Thus, in emergency circumstances, using the food and other supplies already available in subway stations, we can use the subway space as shelter for longer timescales.

Subway stations should have regular as well as independent communications equipment. Communications infrastructures, such as fiber-optic cables, can be built in subway networks. Subways should provide special emergency access and entry and exit points to be used in crisis situations to make rapid population displacement possible. Establishing underground access from the basements of public facilities to the subway can facilitate the transportation of people and protect them in subway shelters and vice versa (transferring population from affected subway stations to

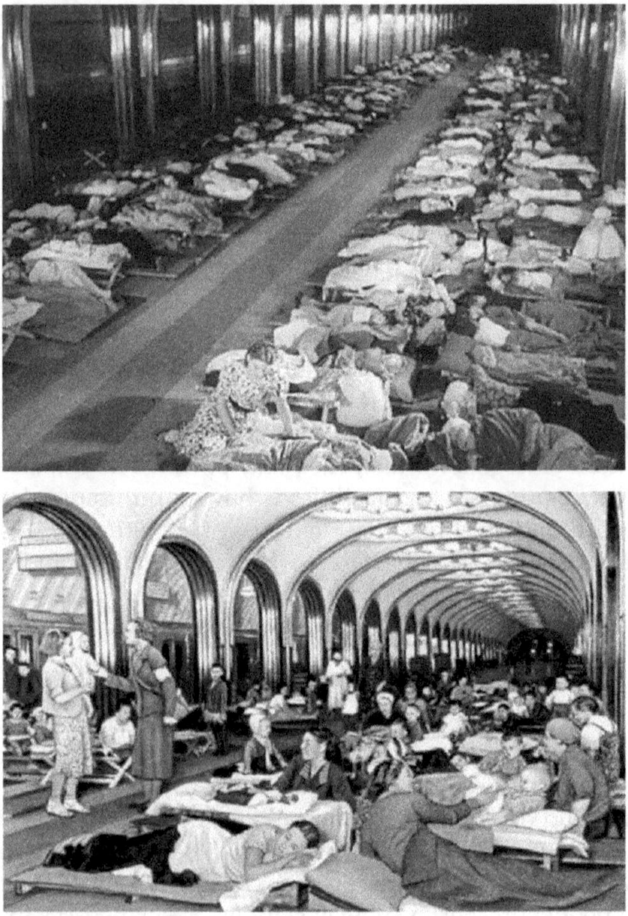

Figure 4.7 Using the Russia subway as a shelter during World War II. (From VK, Presentation on Echoes of the Past War, 2002, http://vk.com/history2worldwar.)

other secure parts of the city). In many developed countries around the world, such as Switzerland, Russia, and Japan, the construction of subway stations is done with special construction measures in terms of material, structures, and building types in order to strengthen them enough to be used as nuclear shelters (Ziari, 2001). Taking advantage of such experiences in equipping and retrofitting existing stations and designing and constructing new stations in our cities is essential for crisis management (Figure 4.8).

Figure 4.8 Antinuclear considerations in the construction of a subway station in Pyongyang, North Korea. (From Bemil, Pyongyang Subway, 2009, http://bemil.chosun.com/nbrd/bbs/view.html?b_bbs_id=10044&num=47253.)

The underground tunnels can also be used in crisis situations to meet different purposes. Subway tunnels can, in times of emergency, be used to transport troops and equipment to allied forces in different parts of the city and create conditions necessary for surprise attacks. Creating diversion tunnels from the subway to special activities and functions at initial stages of drilling and construction can make hidden underground transport and communications possible. These diversion tunnels can be used to connect sensitive areas and security defense uses and create specific functions and strategic activities underground. Overground joints and connection points with underground ones should be created in predetermined locations with inconspicuous forms and structures homogeneous with the surrounding environment. It is worth mentioning that direct and indirect utilization of subway networks should be based on network security considerations. Therefore, the vital, critical, or important uses that are strategic targets themselves, which can be dangerous for the subway due to their specific activities, should never be included. Subways quickly transfer large numbers of people outside urban areas (suburban areas) and thus contribute to urban decentralization and evacuation. Examples of this can be found in the Moscow evacuation plan via subway terminals. In this plan, using Moscow's radial network and the subway's high capacity, people are transferred to predetermined areas in the suburbs in times of disaster. Municipal officials and managers take advantage of Moscow's subway system to reduce harm to civilians (Dai Nejad, 2006) (Figure 4.9).

Airports

Airports are considered a key transportation infrastructure in cities. The role of this infrastructure in times of crisis or war has become more important than ever; if they are destroyed by attacking forces, these cities will incur huge and irreparable financial damage and lose their rapid connections to different key strategic areas. The most important objectives these uses serve in times of war and crisis include quick movement of people, groups, and special forces; relocation of injured and sick people to secure locations with better medical facilities; and transportation of equipment, including weapons and ammunition.

Hence, measures should be taken to protect this infrastructure and the urban texture around it against possible damage and financial loss. Some of the defensive considerations that can be used in order to reduce these injuries and losses are discussed below.

Planning and Design Considerations

- The location of main airports in the city should not be in the path of urban development. In fact, the establishment of airports around the city where further urban development is restricted

Figure 4.9 Moscow subway as a heavy-duty subway can be used for rapid population evacuation and as a well-equipped shelter. (From Transsib, El Metro, 2001, http://www.transsib.com/trans-siberian-city-tours/moscow/moscow-metro.html.)

due to natural and strategic conditions is desired and stabilizes the special airport location in the city.

- Key airports should not be located within cities if possible to prevent damage to the internal physical texture. Therefore, the best location for airports—regardless of the prevailing wind considerations—is on the edge of cities and away from compact physical structures. Indeed, when the airport is away from urban physical textures, the financial and human losses inflicted on urban structures are reduced (Shakibamanesh and Fesharaki, 2011).

- In some cities, airports, especially old ones, are often located in urban areas adjacent to compact physical texture; transferring them out of the area is not possible due to economic, environmental, or military reasons (Shakibamanesh and Fesharaki, 2011). Observing a distinct boundary around the airport site reduces the risk of damage to surrounding buildings. Such boundaries vary depending on the importance of the airport, the physical configuration of the structures, and the importance of potential losses arising from the destruction of its various sections. For example, due to the extent of fires and explosions caused by destruction, the privacy and safety boundary of the flight hangars should be more than that of the airport freight service.
- In major cities, creating one or more small airports (depending on the city scale) as an alternative in case of crisis conditions is necessary. In many cases, losses and damage to the central airport can cut connections of air pathways in that critical period of time. In such circumstances, small airports can be used until the main airport is rebuilt and operational. Old, small airports can be assigned to this purpose, or new ones can be built.
- In enemy attacks to airports, the ground facilities, equipment, and structures are damaged the most. Therefore, correct utilization of underground structures is vital for airport facilities; complementary equipment and hangars can protect them against military attacks and provide the possibility for air defense and protection of the airport site by fighter aircraft (Shakibamanesh and Feshahraki, 2011). The design and location of underground constructions at the airport site should have functional relationships with surface structures and equipment while keeping a convenient and secure distance from each other.
- When establishing airports, care should be exercised so that there are available access roads from different parts of the city; they should be located in the vicinity of major roads with a high level of service so they can be used in war and crisis situations as emergency runways. Major access roads including highways and freeways provide possibilities for such emergency flights. Their presence in the immediate vicinity of the airport is recommended. Also, creating emergency underground access tunnels

can be helpful in connecting hangars and certain parts of the airport with major roads.

- The dispersion and distribution of buildings and physical infrastructure over the airport site should observe the maximum functional relationship between different sectors and related complementary activities, while observing maximum distribution of built structures.
- In many cases, landscape design can be used to distort the correct perception of the airport's macroenvironment, especially from the aerial perspective. This way, enemy forces view the site as split into discrete spaces on land. Moreover, important parts of the site can be camouflaged and some secondary structures can be built with light and cheap materials for peripheral use and deception (Shakibamanesh and Fesharaki, 2011). Vegetation and tree planting alongside the existing topography and ground conditions can help designers landscape for the above-mentioned purpose. Thus, unreal separation and distribution of functional components will make them more difficult for attacking forces to detect and destroy, thereby reducing human and financial losses.
- Independent fire equipment and facilities with proper spatial distribution are essential for airports. When designing interior and exterior spaces, fixed fire extinguishing valves, fire capsules, and intelligent fire extinguishing systems must be established at proper intervals for rapid control of fire.
- In the initial design of airports, special measures should be established to take advantage of auxiliary runways that can be used as alternatives if main runways are damaged in war conditions.
- All natural potentials, including topography, slope, land directions, natural elements, and type and distribution of vegetation on site and in its surroundings, should be taken into account when designing and establishing different parts of the airport, such as hangars, vital facilities and equipment, fuel tanks, and so forth. These measures can greatly reduce losses and make it difficult for attacking forces to identify and destroy sensitive areas.
- Airports should equip shelters to protect the lives of passengers and employees, and thereby assist in crisis and war conditions to reduce casualties.

This chapter began with the question of how can we improve the city resistance in war. In this regard, the necessity of implementing passive defense measures was proposed. According to various sources and published literature mentioned in this chapter, *passive defense* is defined as a set of nonmilitary measures taken to reduce vulnerability and potential damage caused by invading forces, increase resistance of living areas in wars, maintain essential activities in cities and villages, and finally, facilitate crisis management against the military threats and actions. Actually, passive defense is an effort to protect the citizens of a state (generally noncombatants) from military attacks via using the principles of emergency operations, prevention, mitigation, preparation, response, emergency evacuation, and recovery.

Then, this chapter expressed the importance and necessity of passive defense in urban planning and design in terms of major transportation terminals. As it is mentioned, urban transportation terminals play an important role in reducing or increasing the damage and vulnerabilities resulting from attacks. Therefore, decision making when it comes to their locations, structures, and so forth, before their development, as well as solving problems of the current transportation terminals, can make our cities more sustainable from a defensive perspective.

In this chapter, railway, subway, and airport systems are considered critical and valuable urban infrastructures and, in times of crisis and war, play important roles in transporting people, troops, ammunition, weapons, equipment, food, and relief supplies. Since railway, subway, and airport systems are integral parts of today's cities and are part of the spatial structure of cities, they were discussed in this chapter in terms of passive defense. Despite their inflexibility in terms of mobility, some solutions (such as where they are located)—from a specialized urban design perspective rather than a military one—were proposed to reduce the vulnerability of these critical infrastructures and the resulting negative effects exerted on their surrounding areas.

ENDNOTES

1. Vital uses are those whose total or partial destruction will cause crisis and serious damage in the political, leadership, control and commandership, productive and economic, logistics, communication, social, or defensive systems with national effects.
2. Critical uses are those whose total or partial destruction will cause crisis and serious damage in the political, leadership, control and commandership, productive and economic, logistics, communication, social, or defensive systems with regional effects.

3. Important uses are those whose total or partial destruction will cause crisis and serious damage in the political, leadership, control and commandership, productive and economic, logistics, communication, social, or defensive systems with local effects.
4. The appearance of German bombers in the skies over London during the afternoon of September 7, 1940, heralded a tactical shift in Hitler's attempt to subdue Great Britain. During the previous 2 months, the Luftwaffe had targeted Royal Air Force (RAF) airfields and radar stations for destruction in preparation for the German invasion of the island. With invasion plans put on hold and eventually scrapped, Hitler turned his attention to destroying London in an attempt to demoralize the population and force the British to come to terms. At around 4:00 p.m. on that September day, 348 German bombers escorted by 617 fighters blasted London until 6:00 p.m. Two hours later, guided by the fires set by the first assault, a second group of raiders commenced another attack that lasted until 4:30 the following morning. This was the beginning of the Blitz—a period of intense bombing of London and other cities that continued until the following May. For the next consecutive 57 days, London was bombed during either the day or night. Fires consumed many portions of the city. Residents sought shelter wherever they could find it—many fleeing to the Underground stations that sheltered as many as 177,000 people during the night. In the worst single incident, 450 were killed when a bomb destroyed a school being used as an air raid shelter. Londoners and the world were introduced to a new weapon of terror and destruction in the arsenal of twentieth-century warfare. The Blitz ended on May 11, 1941, when Hitler called off the raids in order to move his bombers east in preparation for Germany's invasion of Russia (Johnson, 1990).

REFERENCES

Abbaszadeh Fard, S. (1999). *Reconstruction and Renovation of the War-Torn Areas: Strategies and Policies* [in Persian]. Plan and Budget Organization Press, Tahran, Iran.

American Planning Association (APA). (2006). *Planning and Urban Design Standards.* Wiley Press, Hoboken, NJ.

Army Institute for Professional Development (AIPD). (2008). Conduct smoke operations. U.S. Army Chemical School, Fort McClellan, AL. http://archive.tpub-products.com/content/p/id/44/ (accessed December 12, 2013).

Asghryan Jeddi, A. (1996). Passive defense in Bam citadel [in Persian]. Presented at Proceedings of the Congress of Architecture and Urban Planning of Iran, Bam, Iran.

Baker, S. J. F. (1978). *Enterprise versus Bureaucracy: The Development of Structural Air-Raid Precautions during the 2nd World War.* Oxford, UK: Pergamon Press.

Chairman of the Joint Chiefs of Staff. (2006). *National Military Strategy to Combat Weapons of Mass Destruction*. U.S. Department of Defense, Washington, DC. http://www.defense.gov/pdf/nms-cwmd2006.pdf (December 4, 2011).

Cristy, G. A. (1974). *Technical Directives for the Construction of Private Air Raid Shelters and the 1971 Conception of the Swiss Civil Defense*. Federal Office of Civil Defense, Bern, Switzerland.

Dai Nejad, F. (2006). Principles and guidelines for the design and furnishing of residential open space for passive defense. Housing and Building Research Center, Tehran, Iran.

Iranian Association of Rail Transport Engineering (IARTE). (2014). Iran railway news. http://www.railassociation.ir/html/index.php?op=modload&name=News&file=arview&catid=3 (accessed July 5, 2013).

Johnson, D. (1990). *The London Blitz: The City Ablaze, December 29, 1940*. Stein & Day Pub, New York.

Movahedi Nia, J. (2008). *Fundamentals in Passive Defense* [in Persian]. Malek Ashtar University Press, Tehran, Iran.

Shakibamanesh, A. (2009). Passive defense, public shelters design and planning considerations. Presented at Positioning Planning in Global Crises Conference, Bandong, Indonesia.

Shakibamanesh, A., and H. Fesharaki, S. J. (2011). *Urban Design from Passive Defense Vision* [in Persian]. Boostane Hamid Press, Tehran, Iran.

Ziari, K. (2001). Passive defense and urban shelters planning [in Persian]. *Soffeh Journal*, 11(4), 67.

Ziari, K. (2002). *Urban Land Use Planning* [in Persian]. Yazd University Press. Yazd, Iran.

5

Return to a State of Nature, Compassionate Conservatism, Failed Response, and Their Impact on Race, Ethnicity, and the U.S. Economy

Hurricane Katrina Case Study

Antoinette S. Christophe and Michael O. Adams

Contents

INTRODUCTION

The U.S. government has been aware of the problems with the levees in New Orleans for years. Officials have known that the levees were far below the required limit to provide protection and sustainability for the citizens in the wake of a category 4 or 5 hurricane. The Great Flood of 1927 overwhelmed the lower Mississippi basin. This historic event resulted in the implementation of the Tributaries Flood Control Project of 1928 and the Flood Control Act of 1936. In September 1965, Hurricane Betsy drove storm surge into Lake Pontchartrain, pushing water over levees and flooding the city, resulting in the breaching of the Florida Avenue levee. In the Ninth Ward of New Orleans and Chalmette, floodwaters reached eaves of houses and in some places went over the one-story roofs. The floodwaters overwhelmed the lower Mississippi basin, resulting in 164,000 homes flooding, leaving 675,000 homeless, 74 people drowning, and creating a 30,000-square-mile lake in Louisiana. The Army Corps of Engineer's Hurricane Protection Program exists because of Hurricane Betsy. This country has a record of accomplishment of legislating after the fact, instead of taking preemptive measures to avoid devastation like that caused by Hurricane Katrina.

The United States is a wealthy and technological-based country. However, just beneath that wealth is concealed images of a failed state, the level of a third world country, armed and violent people primed for guerrilla warfare against their neighbors (Purdy, 2005). This was the stage setting in late August–early September 2005. In New Orleans, warning went unheeded, levees were neglected, and cops and rescuers were short-handed or missing. This basic vision captures that of a nation-state, one put forth by Thomas Hobbes four centuries ago, where the people returned to the state of nature; this was New Orleans in 2005 after the devastation of Hurricane Katrina. This only brings to bear what Thomas Hobbes said, that the sovereign has to keep a monopoly of force on the state's own territory via a "Leviathan," or anarchy will result.

Hurricane Katrina, one of the deadliest natural disasters in U.S. history, affected New Orleans, causing extreme devastation to the citizens, infrastructure, environment, and economy. What went wrong in New Orleans after Hurricane Katrina hit and the levees breached? Why did President Bush's response appear to be inept? Why did the left seem uninterested in upholding the Hobbesian responsibility, the Leviathan? Why do Americans, including the impoverished, embrace romantic libertarianism, a doctrine made for neglectful government? How did it influence the environment? What is the resulting infrastructure? When did the port authorities learn of the adverse consequences facing shipping? Much of the answers to these questions are embedded in the ideology of the elite Bush-style conservatism. Much of disaster management may be the product of liberal administrations; however, the state in which the citizens were left in New Orleans that fateful day resulted largely from the compassionate conservative ideology upon which this great nation was built. In this chapter, the foundation from which conservative constitutionalist ideology was born will be the basis for the following considerations:

1. The influence of classical liberalism on America's Constitution
2. The origin of the American institution, as an eighteenth-century product
3. The evolutionary changes that have occurred in the philosophy of government in the United States
4. How conservative constitutionalist ideology can affect the future of the Gulf Coast international trade
5. The relevance of the four aforementioned issues to New Orleans's devastating Hurricane Katrina aftermath and the overarching impact to the economy going forward

This review of literature on the American institution's evolution, idyllic constitutional doctrines, and the administration's response to Hurricane Katrina focuses on these five statements.

THE INFLUENCE OF CLASSICAL LIBERALISM ON AMERICA'S CONSTITUTION

While the feudal system of Britain influenced the creation of the U.S. Constitution and the resulting structure of the American government, the colonists also derived their political heritage from the English experience of governance in the years 1640 and 1688. The Stoic–Jewish–Christian tradition had a caviling effect on the West because it reminded rulers that above their commands there was a higher law, founded on natural reason or divine revelation (Ebenstein and Bernstein, 2000). In 1640, Parliament began asserting its authority, followed by the beheading of Charles I and the appointing of King James II. Then in 1688, King James II began challenging the authority of Parliament by trying to convert England to Catholicism by force. Thomas Hobbes, in 1688, retorted by writing an essay addressing the issue of natural law, saying that civil (or positive) law is derived from and inferior to a higher law, a "law behind the law"—the law of nature. In essence, the law of kings and princes is subordinate to the law of God. This belief was that of the colonist as well. However, King James II met with other challenges; he was branded a traitor and one who sabotaged the Constitution of the kingdom by receiving money from Louis XIV of France. As a result of this betrayal, Parliament, the church, and the army no longer supported James II. Parliamentary supremacy was decisively asserted in the Glorious Revolution of 1688, in which the problem of who was supreme, king or Parliament, was solved. James II left England for France, and Prince William of Orange and his wife Mary were invited by Parliament to become king and queen of England.

The eighteenth-century European philosophy was also grounded in the Enlightenment movement or the Age of Enlightenment, which advocated rationality as a base to the establishment of an authoritative system of ethics, aesthetics, and knowledge. Conversely, the individuals present within this era were more aware of their differences than their similarities, and one key conflict was the role of theology. During the previous period, there had been the splintering of the Catholic Church along doctrinal lines between Catholic and Protestant theologies, which led to a source of partisan debate. Religion was also linked to another feature, which produced a great deal of Enlightenment thought, namely, the rise of the *nation-state*. The movement's intellectual leaders, regarding themselves as courageous and elite, began exploring the question of what constituted the proper relationship of the citizen to the monarch or the state. Their purpose was to lead the world out of a long period of doubtful tradition full of irrationality, superstition, and tyranny, the Dark Ages, toward a world of progress.

Enlightenment philosophers such as Voltaire and Jean-Jacques Rousseau questioned and attacked the existing institutions of both church and state. Voltaire spent 3 years in England, from 1726 to 1729. He wrote *Letters on the English* in 1734, which revealed, to the French, a land of freedom and common sense, secular in outlook, tolerant in religion, and respectful of the rule of law. One contribution Voltaire made unknowingly was opening the way for change by his ruthless and ridiculing attacks on intellectual obscurantism and inequality before the law (Ebenstein and Ebenstein, 2000). Because of his actions, in the second half of the eighteenth century or around the time of the American Revolution, political criticism and philosophical expression became more daring and outspoken. In addition, the focus on law became necessary for the rise of a philosophy that had a very strong concept of the individual. According to this new concept, rights were based on ideals other than ancient traditions; they reflected the intrinsic quality of a person as defined by the philosophers of the age. Among the outstanding philosophers that influenced classical liberalism and the ideology of the American colonist were John Locke, Jean Jacques Rousseau, and Adam Smith.

John Locke, living in England the latter part of the seventeenth century, wrote his *Two Treaties on Government* (1690) to argue that property was not a family right by tenure, but an individual right brought on by mixing labor with the object in question and securing it from other use. This Lockean concept caused the rise of a new class to power in terms of the new political economy and liberated the ingenious or industrious entrepreneur medial labor force. Locke's theory was later used to defend capitalism. Locke also spoke out against absolute monarchy; he believed that in the state of nature, every man had liberty, not just the king. Locke was a believer of a democratic nation, governed by the people. He did not trust the executive power; he had more confidence in the legislature, who represented the people. The text of the Declaration of Independence and the main part of the American Constitution are traceable to Locke, limited government, individual inalienable right, and the inviolability of an individual's property. Both aspects of Locke philosophy, economic and governmental, were used by the American colonists to build this nation and are retained as vital parts of the United States today.

Jean Jacques Rousseau was the author to several essays, but the one that influenced the colonists the most was *The Social Contract*. In the writing, Rousseau is concerned with political obligation and the general will of man. According to him, each man pursues his self-interest in the state of nature until he realizes his power to preserve himself individually against threats and hindrances of other men is not strong enough. At this point, the social contract should protect him with security and liberty. The character of the general will was determined by two elements, the general good itself and whether the general good comes from all and applies to all. The master conception of *The Social Contract* embodies the image of the American colonists after the revolution: a community of free men living in a small state in which democracy can be practiced directly by the people, a community of men who see in freedom not only an invitation to personal enjoyment and advantage, but also shared responsibility for the welfare of the whole (Ebenstein and Ebenstein, 2000). Rousseau is most known for his statement "Man is born free and everywhere he is in chains."

Adam Smith, the apostle of capitalism assumptions, was representative of the British capitalist–utilitarian era. His philosophy concentrated on the importance of free-market, high-quality goods for the lowest price. Smith believed that a free market allowed all individuals in an economy to improve their condition, and moreover, when every individual improves the collective, the nation improves and wealth increases. According to Adam Smith, human beings are autonomous and independent, and it is up to them if they become more or less successful in life. His famous analogy of "the invisible hand" supports this way of thinking: if people are left alone, they will produce not only their greatest good, but also the good for everyone. To Smith, the government's role should be to provide a stable social framework, so that every man could better his condition, without any interference from government. The government role should be limited to functions like dealing with national defense, administration of justice, and facilitating public works and institutions that benefit society. This philosophy would not support the much needed social programs that most of the people left behind in New Orleans post-Katrina benefited from. This viewpoint would also lend itself to assign less worth to this group of individuals. Moreover, Smith's division of labor, another aspect of economic growth, brought about the wealth of nations and personal development, required for the free market to be effective. A free market in labor and capital always directs resources to be used by those who manage them optimally and provide the rewards necessary to encourage innovation and technological advances. Producers that have inefficient processes and practices will lose work, go out of business, and workers will lose their jobs (Ebenstein and Ebenstein, 2000). Also, Smith was opposed to mercantilism due to the restraints it placed on foreign trade, making the nation and the competitor worse off.

Rationalization, standardization, and the search for fundamental unities occupied much of the Enlightenment and its arguments over proper methodology and the nature of understanding. The culminating efforts of the Enlightenment ended with the declaration by Jefferson of inalienable rights overshadowing the idea of divine rights. A few days after the Declaration of Independence was adopted, a committee submitted a plan for a "league of friendship and perpetual Union," but not until a year later, after months of debate, did Congress submit the Articles of Confederation to the states for their approval; the articles did not go into effect until 1781. The American Constitution was the third step in the birth of a new nation. It addressed three compromising issues: (1) large and small states' representation in Congress, (2) North and South and the counting of slaves for taxation and representation, and (3) North and South and the regulation and taxation of foreign commerce. The American Constitution was founded in the ideology of the Age of Enlightenment. This foundation was one of *laissez-faire*, each autonomous, independent man pursuing self-interest with little or no interference or help from the government, while being protected with security and liberty by that government. However, in the end, it is up to that individual to be a successful property owner or not. Those left behind in New Orleans after Hurricane Katrina were not successful property owners.

THE ORIGIN OF THE AMERICAN INSTITUTION, AS AN EIGHTEENTH-CENTURY PRODUCT

Thomas Jefferson drafted a document between June 11 and June 28, 1776, that expressed the convictions in the minds and hearts of the American colonists. The political philosophy of the document was not new; John Locke and the classical philosophers had already expressed its ideals of individual liberty. Jefferson only summarized this philosophy in "self-evident truths" and listed the grievances against the king in order to justify before the world the breaking of ties between the colonies and the mother country. On July 2, 1776, the 13 colonies announced their secession from the British empire and the birth of a new nation, the United States of America. They wanted not only freedom from parliamentary interference, but also severance of all political ties. Then on July 4, 1776, the most famous statement of the revolution was the Declaration of Independence, which encapsulated the ideology of the American Revolution. The revolution ideology that became the American creed is liberalism in its eighteenth- and nineteenth-century meaning, distinct from conservative Toryism, communitarianism, mercantilism, and noble forces prevailing in monarchial church-formed culture.

On November 15, 1777, the Continental Congress adopted the Articles of Confederation, combining the colonies of the American Revolutionary War into a loose confederation. The articles were ratified on March 1, 1781, and revision suggested in May 1787; however, in the end it was determined that the Federal Convention would draft an entirely new frame of government rather than amend the existing articles. Among the chief points at issue were how much power to allow the central government, how many representatives in Congress to allow each state, and how these representatives should be elected—directly by the people or by the state legislators. On September 17, 1787, just 11 years after the 13 colonies announced their secession from the British empire, the founding fathers signed the American Constitution.

The Anti-Federalists opposed the Constitution, and the prospect of establishing a strong central government was unthinkable—it would open the way to tyranny. With the memory of the British violation of civil rights before and during the Revolution fresh in their minds, a bill of rights that would spell out the immunities of individual citizens was demanded. On the other hand, the Federalists favored the Constitution, so a bill of rights was unnecessary. The contemplation was the federal government was limited in its powers and could not interfere with the rights of the people or the states, and that most states had bills of rights. The First Federal Congress took up the question of a bill of rights almost immediately. Congress proposed 12 amendments to the states. Ten of these were added to the Constitution on December 15, 1791, and called the Bill of Rights.

One of the first steps taken after the Constitution was amended was to divide the constitutional authority among the three branches of government. The framers of the U.S. Constitution implemented the new concept of the separation of powers in drafting it. Each branch checks the actions of the others and balances them, which provides that the president, legislators, and judges, although mutually dependent in performing their constitutional functions, are given political, in addition to legal, independence. The president is to be chosen by a group of electors, so that he will have different loyalties and represent different interests from senators, chosen by state legislators; from representatives, directly elected by local constituencies; and from judges, holding office for life and appointed by the president with the consent of the Senate. Moreover, the federal system's Constitution divides governmental powers between the central, or national, government and the constituent (state) governments, giving substantial functions to each. The Constitution delegates legislative, executive, and judicial powers to the national government. It reserves to the state all the powers not granted to the national government and not denied in the Constitution to the state. The concept of separation of powers is also prominent in the state governments of the United States. Due to this federal system structure, local government is given power as first responders, and thus the Robert T. Stafford Act of modern times lacks effectiveness. Furthermore, during Hurricane Katrina, the ineptitude of the public administrators strongly prejudiced the practical man in favor of *laissez-faire*, a divine or scientific harmony between private interest and public advantage. *Laissez-faire* was the notion in New Orleans prior to, during, and after the hurricane. The end result was a lag in disaster response to those most vulnerable. Such was the case for those left behind in the New Orleans Superdome and Convention Center.

THE EVOLUTIONARY CHANGES THAT HAVE OCCURRED IN THE PHILOSOPHY OF GOVERNMENT IN THE UNITED STATES

During the first half of the twentieth century, the American government was concerned with national security. The Sedition Act of 1918 was passed, making it a crime to print, write, or publish any disloyal, profane, or abusive language concerning the form of government of the United States or the American Constitution, or to use language intended to bring contempt, scorn, or disrepute to the U.S. military forces, American flag, or uniforms of the Army or Navy. Then the Smith Act of 1940 was passed, which forbade a person to advocate, teach, or advise forceful overthrow of the government by violence, or to organize any group for this purpose. Later, once the Cold War turned hot in Korea, Congress responded with the Internal Security Act of 1950. This act strengthened laws against espionage and sedition, added to alien registration requirements, made it more difficult for Communist aliens to enter or remain in the United States, and established procedures for detaining, in the event of a national emergency, any person who could reasonably be expected to engage in acts of sabotage or espionage. This series of legislation was finalized with the Communist Control Act of 1954, which deprived the Communist Party and its successors of "any of the rights, privileges, and immunities created under the laws of the United States" or any of the states. This was the first time in history that the national government outlawed a political party and denied citizens the right to use the traditional instruments of democracy to achieve their political goals (Burns and Peltason, 1972).

In the late nineteenth and early twentieth centuries, socialism as an organized political force made its appearance. People of government were concerned that the two opposing economic philosophies would place a strain on democracy. Capitalism moved in the direction of socialism with respect to government provision of services and macromanagement of the economy. This was the result of democratic idealism of human dignity being incompatible with personal insecurities, social inequalities, economic want, and human suffering when there were means available to remedy such conditions. However, the socialists moved in the direction of a classical political economy on the issue of private ownership because productive resources through nationalization proved inefficient, and more importantly, the democratic values of power's diffusion and personal liberty were incompatible with total ownership and management of the nation's economy by the state (Ebenstein and Ebenstein, 2000). As social discourse developed during the twentieth century, there was increasing focus on economic, as opposed to strictly political, issues. At least two aspects of the welfare state had emerged in the industrial democracies during the twentieth century: provision of social welfare and national economic management or intervention (Ebenstein and Ebenstein, 2000). John Maynard Keynes, a British economist and elitist, played a major role in this era. He believed that government should manage, rather than control, the economic conditions in which private enterprise takes place. In Keynes book *A Tract on Monetary Reform* (1923), his greatest concern at the time was that domestic prices be stabilized from inflation or deflation, rather than the primary goal of monetary policy to fix intercurrency exchange rates. He felt government should control money supply so as to level economic expansion and contraction, not to react to changes in currency exchange rates.

Around this time, the country was at war and national security was an issue. During World War I around 1918, blacks began to migrate to northern cities, and by then there were small beginnings toward educational opportunities and jobs. The president had appointed federal judges more sympathetic to the construction of the Thirteenth, Fourteenth, and Fifteenth Amendments—judges who intended to impose the amendments. In the 1930s, around the time of the Great Depression, unemployment was prevalent and national incomes dropped by a third. Social security (1935) was implemented in the United States under Franklin Roosevelt's New Deal, and later was expanded into a program that was not tied to paid taxes. The program was the Aid to Families with Dependent Children (AFDC). The program was originally intended for widows with small children (from World War I), but over time it grew to include mostly unmarried women with children and was supplemented by various other benefits in such areas as nutrition, housing, and medical services. The welfare state was a new way of organizing a society, one that would be ground in the fabric of twenty-first-century New Orleans's disadvantaged population known as the Ninth Ward and New Orleans

East. The poor, elderly, unemployed, and infirm would not have to rely on voluntary donations or private charity; instead, they would have a legal right to support from the national government. From the 1940s to 1960s, most national governments in the Western world used Keynesian macroeconomic policies. During this same time, blacks started to resort to litigation to secure their rights, challenging the doctrine of segregation and discrimination.

The unfolding of a welfare state was bound to cause a reaction in the U.S. economy. The encounter of capitalism with socialism resulted in a mixed economy, a welfare state that seeks to combine the best of capitalism and socialism. The gross national product (GNP) went from 2.5% in 1926 to 20% in 2000, an eightfold increase. The local, state, and national spending increased as well. The scope of government, as well as the size, had enormously increased. By the 1960s, the nation was beginning to embrace libertarianism, a move toward less government. This was a conservative reaction to the welfare state and general growth of government. Friedrich Hayek and Milton Friedman were leading academic proponents of less government during the twentieth century. Hayek, an intellectual inspiration for many in the Reagan and Thatcher governments, wrote *Law, Legislation, and Liberty*, in which the ultimate goal was a world of universal law and a world federation of free nations. Friedman agreed that there was a substantial role for government in establishing the parameters of the nation's economy, but he disagreed and had great influence on the reinterpretation of the cause of the Great Depression. Friedman's analysis revealed that the cause of the Great Depression was the nation's monetary mismanagement, which could have been corrected. This concept has become the mainstream view within academic economics.

Most of the Western nations were experiencing high inflation and high unemployment toward the end of the 1960s and early 1970s. This was called stagflation. Standard Keynesian analysis could not account for stagflation. Friedman's position was that a free market almost invariably will provide better services cheaper than government, and that the government should stay out of the way. From the 1960s through the 1990s, Friedman was the leading popular and academic factor of libertarianism. The welfare state eliminated a lot of deprivation among intact families and the elderly and created significant wealth, direction, and ownership by private individuals. The purpose of the welfare state was to better people's conditions and produce economic growth.

However, in the 1980s and 1990s, there was a shift to the right, with retrenchment of the welfare state, a backlash against "tax-and-spend liberals," and the election of conservatives (the New Right Party). Most supporters of the New Right Party acknowledge some obligation to redress unequal opportunities and protect the vulnerable, but they had mixed emotions. They didn't reject the liberal egalitarian goal of an ambition-sensitive, endowment-sensitive distribution, but instead thought that the welfare state had failed to achieve any of its goals. They promoted that the liberal egalitarian redistribution policies wouldn't enable the disadvantaged to enter mainstream society to exercise their civil and political rights, but would promote a passive nature among the poor, create cultural dependency of the impoverished, and without actually improving their life chances. Furthermore, the welfare state was seen as taxing hard-working citizens to subsidize the lazy or indolent who simply do not want to work, which, in their mindset, violates the norm of ambition sensitivity and the principle that people should be responsible for their own choices. The public opinion polls confirm that people today are more likely than those 20 years ago to say that people on unemployment insurance or welfare benefits are responsible for their condition, rather than being a victim of circumstance, misfortune, or unequal opportunities. Many thought this way of the people that sought refuge in the New Orleans Superdome and Convention Center post-Katrina.

The 1990s brought a new president, Bill Clinton (1993–2001), a cautious, moderate Democrat that joined Republicans in reducing welfare benefits and the size of the federal workforce. In 1996, Clinton conferred with the Republican-controlled Congress to put into practice the Personal Responsibility and Work Opportunity Act. This new piece of legislation radically reorganizes AFDC program. The revised AFDC program, called Temporary Assistance for Needy Families (TANF), enforced a lifetime limit of 5 years for those that received these benefits. The purposeful legislative action was designed to get people out of the cycle of poverty and into the workforce, but for some, it became an unfair denial of aid and a pipeline to homelessness and no way out. Such was the case for many impacted by Hurricane Katrina, who died, watched loved ones overcome by the flood, or simply experienced hopelessness while trapped in the New Orleans Superdome and Convention Center.

However, the government continued to play a crucial role in the nation's economy, with most of the major innovations of the New Deal and the Great Society remaining in place and the Federal Reserve System continuing to regulate the overall pace of economic activity, while keeping a watchful eye on renewed inflation. Americans ended the 1990s with a restored sense of confidence. By the end of 1999, the economy had grown continuously since March 1991, the longest peacetime economic expansion in history. Unemployment totaled just 4.1% of the labor force in November 1999, the lowest rate in nearly 30 years. In addition, consumer prices, which rose just 1.6% in 1998 (the smallest increase except for one year since 1964), climbed only somewhat faster in 1999 (2.4% through October). Many challenges lay ahead, but the nation had weathered the twentieth century and the enormous changes it brought.

Politically and economically, blacks have made substantial strides in the post–civil rights era. Civil rights leader Jesse Jackson, who ran for the Democratic Party's presidential nomination in 1984 and 1988, brought unprecedented support and leverage to blacks in politics. In 1989, Virginia became the first state in U.S. history to elect a black governor, Douglas Wilder. In 1992, Carol Moseley-Braun of Illinois became the first black woman elected to the U.S. Senate. There were 8,936 black officeholders in the United States in 2000, showing a net increase of 7,467 since 1970. In 2001, there were 484 mayors and 38 members of Congress. The Congressional Black Caucus serves as a political bloc in Congress for issues relating to African Americans. The appointment of blacks to high federal offices—including General Colin Powell, chairman of the U.S. Armed Forces Joint Chiefs of Staff, 1989–1993, and U.S. secretary of state, 2001–2005; Dr. Condoleezza Rice, assistant to the president for national security affairs, 2001–2004, and confirmed secretary of state in January 2005; Dr. Ron Brown, secretary of commerce, 1993–1996; and Supreme Court justice Clarence Thomas—also demonstrates the increasing visibility of blacks in the political arena. This most recently culminated in the election of the United States' first African American president, Barack Obama (2009–2016).

KATRINA: NEW ORLEANS'S DEVASTATING AFTERMATH

Today, the Hurricane Katrina disaster has exposed the overwhelming problems with contemporary urban liberalism. This devastating event resulted from two failed political philosophies: a national conservatism unconcerned about urban centers and an urban liberalism unconcerned about the daily realities of the majority of urban dwellers. The overriding cause of the nation's urban catastrophes is modern liberal social policies. Big cities, being the political stronghold of liberalism, were supposed to be the laboratories for the Great Society expansion of the welfare state. Instead, what they became were principal victims of a liberal ideology.

There are three more failures of liberalism, which, when merged, produce a ruinous urban policy. The first is the failure to nurture the sources of economic growth. Second, failure is due to the administration's inability to understand urban neighborhoods. Lastly, is it the failure of the system to appreciate the importance of a strong moral order.

In big cities like New Orleans, the governmental economy creates first-class tourist attractions, but they also produce an environment of third-world inequalities. This misgiving is rooted in the conception of the leaders of the eighteenth century. The authors of the U.S. Constitution were individuals who came to America with nothing and were left behind to build this great nation; they were the product of a mercantile system and the embodiment of *laissez-faire*. Philosophers of that time believed man's labor created and determined the value of his property. Man depended on himself and the help of the community, not the government. Our Constitution was extracted from this worldview. President George W. Bush's ideology as a constitutionalist was bolstered from this environment.

George W. Bush, the compassionate conservative, saw the social problems of the United States, such as healthcare, immigration, and the environment, as being issues that conservatives could find solutions to better than repairing failed programs of liberals and the war on poverty. Compassionate conservative philosophy argued for the return of stigmatism for illegitimacy to encourage two-parent families, welfare reform including workfare, active policing, standard-based schools, and helping poor countries around the world. The administration thought compassionately to actively help our citizens in need, but the conservative side insists on accountability and results. Which was the dominant side on August 29, 2005? The White House claims it aims for a society that helps others to help themselves, thereby avoiding their reliance on the government. Was this the mindset days after Hurricane Katrina hit New Orleans?

Government officials placed the future of New Orleans on a gamble that the arts, nightlife, and tourist economy would build prosperity. The problem with this is that while great restaurants, music, and the nightlife might appeal to the urban rich, the economics of tourism leave huge segments of the population behind. The mayor of New Orleans and other officials relied on the elite culture to fuel their economy, but instead, the results unparalleled class disparity. New Orleans's population was made up of about 28% whites, who were wealthier and more educated than whites nationwide. This group loves the new nightclubs, coffee shops, and the Big Easy's upscale neighborhoods. Surrounding all this, nonetheless, is one of America's most impoverished black communities. About 40%

of New Orleans households, mostly blacks, earn less than $20,000 a year, twice the proportion in the rest of the country. Just like on August 29, 2005, whatever might have worked well for Garden District elites failed the rest of the city's population on a staggering scale. Inequality took its highest toll that day in New Orleans.

The African American middle and professional classes have grown in recent decades, but the wealth of black families in the United States remains on average only 1/10 that of whites. Families in poor, black neighborhoods, like those of New Orleans, often live on monthly checks, with few or no assets. Katrina came at the wrong time of the month, when cash was scarce or gone. The 2000 census reveals that 27% of New Orleans households (~120,000 people) did not have privately owned transportation. The evacuation order came on August 28, 2005, to people who had no way to leave or, if they found a way out, would have been stranded elsewhere with no money and no place to stay. Three days of supplies for 15,000 citizens was delivered to the Superdome by the Federal Emergency Management Agency (FEMA), but a much larger crowd had shown up seeking refuge. By the time the federal government and much needed supplies arrived on Friday, September 2, 2005, the Leviathan had been replaced with anarchy. The basis of this libertarian indifference is denial that in a complex society, private security and private virtue ultimately depend on the state's monopoly over violence. Private life, generous, imaginative, and free, is conducted against the backdrop of state power—the power that enforces private contracts, distributes private property, and will jail or even kill the intruder who tries to force his way into a private home. Without that security, people become dangerous to one another, not because most people are predatory, but because some are, and in a world without the Leviathan, paranoia and preemptive violence grow.

Romantic libertarianism is a doctrine made for neglectful government. It was the doctrine of George W. Bush's Republican Party. The state exists in part to do what private citizens generally will not and cannot do: plan and execute collective action to avert faraway and uncertain disaster. The danger to New Orleans from a hurricane strike was familiar. From a capitalistic libertarian perspective, New Orleans was a poor, black, Democratic city with a failing economy; it was never going to be high on this administration's list for infrastructure development or disaster relief planning. President Bush coining the phrase "compassionate conservative" just offered a new way of thinking about the poor. Paralyzing the poor with thoughts of their own helplessness and inadequacy is completely against Bush's doctrine. He would most likely believe the poor need the larger society's moral support;

they need to hear the message of personal responsibility and self-reliance, the optimistic assurance that if they try, they will make it. They need to also know that they can't blame "the system" for their own wrongdoing—it's not the American way. Much of this paradox and the modification of the AFDC resulted in a "return to a state of nature" that fateful day when Hurricane Katrina struck and the levees failed.

GLOBAL WARMING, HURRICANES, AND INTERNATIONAL TRADE

As an addendum, for at least 200 years the Port of New Orleans, another source of city revenue, has been an important part of America's economy. Easy access to the mouth of the Mississippi River was the principal reason that President Thomas Jefferson spearheaded the Louisiana Purchase in 1803. Prior to Hurricane Katrina, 62% of America's consumption cargo was shipped through the Port of New Orleans, carrying steel, rubber, coffee, plywood, and other consumables (EconSouth, 2005). See Table 5.1 for a description of economic performance prior to Hurrican Katrina. However, immediately after the devastation caused by Hurricane Katrina, the Port of New Orleans was forced to route cargo to other ports, like Jacksonville and Mobile. The Port of New Orleans's desire was to work up to pre-Katrina trading. However, on June 5, 2008, the assistant secretary of the Army for civil works sent the final Mississippi River Gulf Outlet (MRGO) deep-draft deauthorization report to Congress, ending navigation between the Gulf Intracoastal Waterway (GIWW) and the Gulf of Mexico (Department of the Army, 2008). After deep-draft deauthorization of MRGO, plans were put in place to construct a rock closure structure at the south ridge of Bayou La Loutre in St. Bernard Parish. It is anticipated that as a result of MRGO's closure, coastal erosion of Louisiana will be minimized or reversed, and restocking of the St. Bernard Parish and Lower Ninth Ward waterways with their normal aquatic habitat will only improve the coastal and aquatic environments, provide restoration of a natural barrier to hurricanes, and ultimately improve the quality of life for the people in the area.

Today, the Port of New Orleans continues to be central to the busiest global port complex, which has at its disposal a multitude of movement alternatives. Due to all this commerce, as stated by the Port of New Orleans president, "Business Facilities Magazine and Forbes Magazine have ranked the Port as the number 1 logistics hub in the US" (LaGrange, 2014).

Table 5.1 Port of New Orleans Economic Analysis Pre–Hurricane Katrina

Job provided by the port	107,345 (2001) → 160,498 (2004)
Job personal income (average salary $42,000)	$2.3 million (2001) → $8.5 million (2004)
Marine cargo and vessel activity	$17.8 billion in total economic activity (Louisiana)
Revenue received by businesses providing cargo handling vessel service and inland transportation	Total $1.4 billion ($228.4 million in-state purchase; $513.6 million salaries paid direct to employed local residents)
Export growth outpaced import growth (Huether, 2005)	Export 8.3%; import 5.5%

On November 8, 2013, President Obama affirmatively stated, "This [New Orleans] is one of the busiest port complexes in the entire world... the U.S. ports need to be capable of handling the larger vessels the expanded Panama Canal will accommodate or risk losing shipping trade to other countries" (Hand, 2013). Nevertheless, as stated by Parsons (2014), "While loss of life and economic damage from Katrina was large, the effect on the trade, and indeed the overall economy was small." Needless to say, citizens of one of America's most impoverished black communities were not employed by the Port of New Orleans and were probably least affected by MRGO's closure. However, the closure of MRGO was most likely the best decision due to global warming (warm water) and the instability in the lower atmosphere that creates the energy source of hurricanes. With this premise in mind, many companies may become weary of the consequence of trading in the Gulf Coast region, and while it might be a little

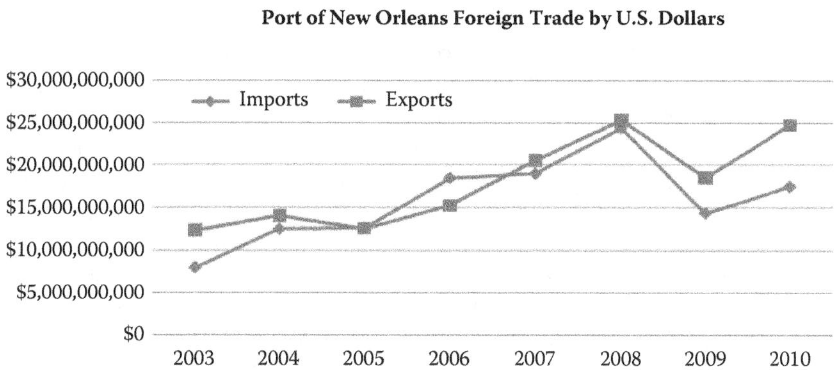

(Source: Foreign Trade Division of the U.S. Census Bureau and USA Trade Online.)

costly to the companies, they will simply relocate to other ports when necessary, and it will be business as usual. This will impact the regional economy, but it most probably will not influence the national economy or international trade market.

CONCLUSION

The American institution, as an eighteenth-century product, grew out of a British feudal system that was instituted by individuals fleeing England in the hope of being able to practice freedom of religion. Thomas Jefferson's philosophy of self-evident truths was influenced by Locke's, Rousseau's, and Smith's classical philosophies. Jefferson's viewpoint was embraced by the founding fathers, who signed the American Constitution, which divided constitutional authority among the three branches of government, and its Bill of Rights ensured the immunity of individual citizens.

In the nineteenth and early twentieth centuries, capitalism, socialism, and classical political economy, being in constant flux, birthed two aspects of the welfare state: provision of social welfare and national economic management or intervention. AFDC was the product of social welfare in the 1930s, and TANF resulted from national economic management in the 1990s. This governmental design was seen as taxing hard-working citizens to subsidize the lazy or indolent who simply do not want to work and, as a result, are responsible for their own choices. This compassionate conservatism has subsisted over the past couple of centuries. The thought process of many current leaders and legislators is swayed by the right-wing political viewpoint, a foundation of *laissez-faire*, each autonomous, independent man pursuing self-interest with little or no interference or help from the government, while being protected with security and liberty by that government. The welfare state was a new way of organizing a society, one that would be ground in the fabric of twenty-first-century New Orleans's disadvantaged population known as the Ninth Ward and New Orleans East—a big city whose governmental economy was created on the back of first-class tourist attractions, but produced an environment of third-world inequalities. This misgiving is rooted in the conception of the leaders of the eighteenth century all the way to modern-day officials.

This philosophy would not support the much needed social programs that most of the people left behind in New Orleans post-Katrina benefited from and would promote devaluation of that group self-worth, and possibly the support of the government for security and liberty. The consequences of this type of philosophy were evident for communities such as the Ninth Ward, New Orleans East, and the Desire Projects of New Orleans in the days after Hurricane Katrina made landfall, August 2005. President Bush, coining the phrase "compassionate conservative," just offered another way of thinking about the poor. Paralyzing the poor with thoughts of their own helplessness and inadequacy is completely against Bush's doctrine. He would most likely believe the poor need the larger society's moral support; they need to hear the message of personal responsibility and self-reliance, the optimistic assurance that if they try, they will make it. They need to also know that they can't blame the system for their own wrongdoing—it's not the American way.

This was the environment, along with the negative impacts of global warming, in which the Port of New Orleans trade operations were conducted and the future immensely altered. Hurricane Katrina made landfall as a category 4 storm on August 28, 2005. One day later, 80% of the city of New Orleans was flooded, with some areas being under as much as 20 feet of water. The overflow of the Gulf of Mexico into New Orleans was indirectly contributed to Hurricane Katrina; actually, it resulted from several levee breaches. The breaches occurred due to a number of factors: powerful storm surge, strong wind, excess water levels in the bodies of water surrounding the city, and poor structural design of the levees. Research shows that as a result of global warming, the earth's temperature is increasing ~3°C per decade. Because of this continuous warming trend, scientists predict that the frequency and severity of cyclonic activity in the Atlantic Ocean and the Gulf of Mexico will steadily rise.

International trade is an exchange in which major disaster events, such as Hurricane Katrina, can have an extraordinary effect on the U.S. national economy. The economic effects of Hurricane Katrina were very devastating to the Gulf Coast; the impact on the U.S. economy for the third quarter of 2005 showed a 26% slowing of economic growth, with a modest negative fourth quarter, and an accelerated economic growth rate in the first half of 2006.

In light of these findings, the Port of New Orleans is very important to the U.S. economy and has some impact on the New Orleans economy, but it does not serve as a major employer for those that were stranded at the Superdome and Convention Center post–Hurricane Katrina. More important is the deauthorization of MRGO. This legislative action is anticipated to curb wetland loss, decrease salinity and natural habitat changes, restore the aquatic life, and curb the potential flooding in St. Bernard Parish. It will also end the chances of devastating impact to business located in a vulnerable location for deep-sea navigation.

REFERENCES

Bea, K. (2007). *Federal Emergency Management Policy Changes after Hurricane Katrina: A Summary of Statutory Provisions.* Congressional Research Service (CRS) Report 7-5700 for Congress.

Blonigen, B. A., and Wilson, W. W. (2008). Port efficiency and trade flow. *Review of International Economics* 16(1), 21–36.

Burns, J. M., and Peltason, J. W. (1972). *Government by the People,* 4–142. 8th ed. Upper Saddle River, NJ: Prentice-Hall.

Christophe, A. (2006). Environmental impact of Hurricane Katrina on international Trade. Presented at SSSA Conference San Antonio, Texas.

Department of the Army, Office of the Assistant Secretary, Civil Works. (2008). Letter from John Paul Woodley to James L. Oberstar, Chairman of the Committee on Transportation and Infrastructure.

Department of the Army, Office of the Chief of Engineers. (2008, January 29). *Mississippi River Gulf Outlet, St. Bernard Parish, Louisiana, Deep Draft Deauthorization Study.* CECW-MVD 1105-2-10a.

EconSouth. (2005). Each week has gotten better: An interview with Gary LaGrange. http://www.frbatlanta.org.

Ebenstein, W., and Ebenstein, A. (2000). *Great Political Thinkers: Plato to the Present,* 355–503, 785–805, 814–824. 6th ed. Belmont, CA: Thomson-Wadsworth.

Foreign Trade Division of the U.S. Census Bureau and USA Trade Online. (n.d.). Retrieved August 7, 2014, from http://www.worldportsource.com/trade/USA_LA_Port_of_New_Orleans_254.php.

Hand, M. (2013, November 11). New Orleans pushes for dredging during Obama visit. Seatrade Communications Limited. Retrieved from http://www.seatrade-global.com/news/americas/new-orleans-pushes-for-dregding-during-obama-visit.html,

Huether, D. (2005). *Hearing on "Commerce and Consumer Protection Implications of Hurricane Katrina."* Washington, DC: National Association of Manufacturers.

LaGrange, G. (2014, July 11). Port profile: The Port of New Orleans. *Port Vision News.* Retrieved from http://www.portvision.com/news-events/press-releases-news/port-profilethe-port-of-new-orleans.

Parsons, C. R. (2014). Disasters and trade: Did Hurricane Katrina affect U.S. imports? *International Economics and Economic Policy,* 12(1). http://link.springer.com/article/10.1007/s10368-014-0296-5#.

Plume, J. (2005). Starting over: The Port of Gulfport is at a crossroads in its post-Katrina reconstruction. *Journal of Commerce.* http://www.bdpcustomercorrect.com/news/KatrinabattersGulfports.asp

Port of New Orleans. (2014). Port statistics. Retrieved August 5, 2014, from http://www.portno.com.

Purdy, J. (2005). Democracy and disaster. http://www.newamerica.net.

U.S. Army Corps of Engineers. (2012). Mississippi River Gulf Outlet (MRGO) ecosystem restoration plan: Orleans, St. Bernard, and St. Tammany Parishes, Louisiana. Retrieved June 14, 2012, from http://www.usace.army.mil/Portals/2/docs/civilworks/CWRB/mrgo_eco/mrgo...

U.S. Army Corps of Engineers. (n.d.). New Orleans district. Retrieved September 1, 2014, from http://www.mvn.usace.army.mil/Media/NewsReleases/tabid/9286/Year/2009/Default.aspx.

Section III

Resilience, Cooperation, and Citizen Attitudes

6

The Big Spill

Who Was to Blame and How Should Government Respond? Citizen Attitudes in the Aftermath of the Deepwater Horizon Spill in Coastal Alabama

Michael Howell-Moroney and Kent R. Kerley

Contents

INTRODUCTION

On April 20, 2010, a major explosion and fire occurred on Transocean's drilling rig, *Deepwater Horizon*, which resulted in the deaths of 11 workers. The rig was licensed to oil giant British Petroleum (BP) for its Macondo Prospect project taking place about 40 miles southeast of Venice, Louisiana. The rig sank in approximately 5,000 feet of water and was 13,000 feet under the seabed. Following the explosion and fire, crude oil was discharged from the well at a rate of up to 5,000 barrels per day. The major companies involved—BP, Transocean, and Halliburton—would attempt frantically over the next several weeks to cap the well. They were not able to do so until July 15, and the well was not completely disabled until September 10. The end result was 4–5 million barrels (200–210 million gallons) of crude oil spilled along the Gulf Coast states of Alabama, Mississippi, and Louisiana (Freudenberg and Gramling, 2012; Reuters, 2010; Weber, 2010). Years after this human-made disaster, the environmental, economic, and social consequences are still being discovered.

In this chapter we explore the associations among event framing, accountability, risk, and policy preference using the case of the *Deepwater Horizon* oil spill. Framing and blaming are two important heuristic lenses through which disasters and risks are viewed. These lenses themselves are socially and culturally determined and can be malleable. The social amplification of risk framework (SARF) provides insight into how these heuristic lenses can magnify or attenuate risks. Amplified risks often result in large secondary impacts, human responses to disasters, or risk events. Government responses to disasters are a key example of secondary impact. We focus on two primary research questions:

1. How does event framing affect perceptions of accountability and risk, and how does it affect policy demands?
2. How do perceptions of accountability affect perceptions of risk and policy demands?

BACKGROUND OF THE *DEEPWATER HORIZON* SPILL

Once considered BP's invincible offshore oil drilling rig, *Deepwater Horizon* was lying on the ocean floor in the spring of 2010. Estimates of the breadth of the spill were shocking to most. The best estimates are that 4–5 million barrels, or 200–210 million gallons, of crude oil were spilled along the Alabama, Mississippi, and Louisiana Gulf Coast states (Freudenberg and Gramling, 2012). In terms of blame for the incident by those affected (the general public, researchers, and media members), BP appeared to be the primary culprit. Disaster scholars, for example, attributed the incident to "a series of cost-cutting moves" taken by BP and its contractors (Freudenberg and Gramling, 2012).

Although the bulk of early media reports contained depictions of oil spillage on the Louisiana and Mississippi Gulf Coasts, Alabama had approximately 50 miles of coastline decimated by the *Deepwater Horizon* spill. The Alabama Gulf Coast is comprised of two large counties (Baldwin and Mobile), and at the time of the spill, there were over 600,000 residents in those counties alone. It is nearly impossible to estimate the economic impact to the state given that the bulk of losses were in the tourism sector. Estimates of recovery expenses, as well as current and future revenues lost, range from several hundred million dollars to $3.3 billion (Addy and Ijaz, 2010).

Even with sizable settlement funds from BP, it is unclear among those who study marine and aquatic disasters how much the affected areas will be able to recover. Freudenberg and Gramling (2012) estimate that only 5%–10% of the areas can be cleaned due to the boundless nature of the sea. Not surprisingly, then, the *Deepwater Horizon* incident is understood by most as a human-made disaster with long-term environmental, economic, and social consequences.

As this chapter was being prepared, a federal judge handed down a major ruling concerning the *Deepwater Horizon* spill. On September 4, 2014, U.S. District Court judge Carl Barbier issued a ruling of "gross negligence" in regards to BP's role in the spill. Judge Barbier ruled that the drilling rig company Transocean and oil services company Halliburton were also "negligent," but that 67% of the fault was with BP. Under this ruling, BP will be subject to fines of up to $4,300 per barrel spilled, which is in accordance with the Clean Water Act as amended in 2002. Monies received from those fines will be distributed to those in the affected states via settlement fund administrators. Executives from BP claimed to have paid out approximately $28 billion in fines already, but this latest ruling creates the possibility of up to another $18 billion in fines based on estimates of 4.2 million barrels spilled (Oliver, 2014).

On the Alabama Gulf Coast, recent estimates are that well over 30,000 businesses and individuals have filed claims and received payments from BP via the state's settlement fund administrator. It is estimated that those claims resulted in payouts to date of close to $600 million (Diel, 2013). The state will be eligible for its portion of the new funds as part of the September 4 ruling (minus administrative expenses from federal and state governments).

SOCIAL AMPLIFICATION OF RISK FRAMEWORK

Disasters like the BP oil spill create a number of impacts. Primary impacts resulting from a disaster are those linked most directly to the event. In the case of the BP oil spill, examples of primary impacts include loss of human life, oil incursion into marshes and beaches, and damage to marine and aquatic life. But disasters often also have ripple effects beyond their primary effect. They often create a series of secondary impacts that result from human reaction to the event. Formal governmental responses in the form of new policy or regulation are important secondary impacts that often result from a disaster.

The social amplification of risk framework is a theoretical basis for understanding how disaster events trigger (or fail to trigger) secondary impacts. Although risk is often portrayed as an objective quantity that can be measured precisely, the SARF considers the cultural and social nature of risk perception. Born out of communications theory, the SARF posits that disasters and their attendant risks are framed and interpreted through a variety of "stations." These stations serve to amplify risk signals or attenuate them. In this sense, the SARF proposes that risk is a socially determined concept that is itself a result of a heuristic interpretation of events (Douglas, 1992; Kasperson et al., 2003; Kasperson et al., 2005). Heuristic interpretation of risk through social and cultural channels is the major mechanism through which human response to risk events is mediated.

When risks are greatly amplified, this results in dramatic and far-reaching policy change over a short period of time. For example, the anthrax postal scare in 2001 led to a large and costly federal program to detect anthrax in postal facilities. On the other hand, some behaviors, such as driving, are highly risky, but those risks are attenuated by a series of stations that diminish the perceived risk from the activity. Other risks are diminished because they are the result of so-called crescive problems, which incubate over time and go largely unnoticed (Beamish, 2002). From the standpoint of public policy, the SARF is a conceptual framework for understanding how different factors serve to amplify or attenuate citizen perceptions of risk. It is those mechanisms of amplification and attenuation that ultimately influence institutional responses to risks in the form of public policy (Freudenberg, 2003).

Framing and Blaming

According to the SARF, perceived risks are an amalgam of actual risks and other malleable perceptions. This heuristic component of risk perception means that it can be amplified or attenuated; that is, risk is perceived and framed through specific perceptions and "stories" surrounding events. Event framing plays a central role in human understanding and interpreting events. Framing provides a heuristic lens for us to understand how and why an event occurred. It also leaves us with impressions about the likelihood of an event recurring at some point in the future (Pidgeon, 1997; Pidgeon and O'Leary, 2000). Kingdon's (1985) classic work on event framing reminded us that perceptions of problems or events also govern the set of solutions that are considered.

In the case of the BP oil spill, one important frame to consider is the degree to which people believe that the spill was an isolated incident or a symptom of a larger underlying risk attributable to the activity of offshore drilling in general (Hoffman and Jennings, 2011). This frame has important implications for perceptions of both risk and policy preference (Bozeman, 2011). For example, if a person perceives the BP spill to be an isolated incident, then perceptions of risk from future oil spills likely will be low. On the other hand, people who view the spill as merely a symptom of larger systematic risks associated with offshore drilling will likely perceive higher risks of future spills.

Event framing also carries over into the policy realm. If a higher level of risk is perceived with a given activity, this likely will translate into calls for greater regulation, or perhaps even banning the activity. Consider the conclusion reached by the National Commission on the BP *Deepwater Horizon* Oil Spill and Offshore Drilling in 2011:

> In short, the safety risks had dramatically increased with the shift to the Gulf's deep waters, but Presidents, members of Congress, and agency leadership had become preoccupied for decades with the enormous revenues generated by such drilling rather than focused on ensuring its safety. With the benefit of hindsight, the only question had become not whether an accident would happen, but when. On April 20, 2010, that question was answered. (p. 85)

On the other hand, if an event is viewed as isolated, the justification for large-scale regulation often is not present. Instead, the appropriate response to the incident can be as narrow as punishing the guilty parties for negligence. For example, the conclusions of the commission were contested by an oil industry lobbyist:

> API Upstream Director Erik Milito said the group is still in the process of reviewing the commission's report but is pleased the commission is recommending increased funding for the federal agency responsible for inspecting and monitoring offshore activity. However, he said API is deeply concerned that the commission's report casts doubt on an entire industry based on its study of a single incident. "This does a great disservice to the thousands of men and women who work in the industry and have the highest personal and professional commitment to safety," Milito said. (Palomo, 2011)

Institutional blame and trust are concepts closely related to event framing (Frewer, 2003). The question here is which institutions were held responsible in the postmortem dissection of particular events. Oil spills, in particular, are uniquely suited to the "blame game." Gochfeld and Burger (1994) propose that "oil spills are... unlike natural disasters such as earthquakes and volcanoes, there is usually no loss of human life and there is always a human agency to blame" (p. 275). The dynamics of institutional accountability for disasters create a special sort of event frame because they draw the causal lines of responsibility for a given event to certain actors. These lines of responsibility, in turn, lead to an attendant set of decision rules for interpreting the facts of a given disaster. Douglas (1992) writes, "Blaming is a way of manning the gates through which all information has to pass" (p. 19).

While ascribing accountability to BP for the spill seems straightforward enough, federal government accountability requires a more nuanced view. We argue that persons who perceived that the federal government was accountable do so because they believed that the federal government should regulate offshore drilling. In this way, federal accountability existed because the spill happened as a result of lax regulation or oversight.

THE ALABAMA COASTAL CITIZEN SURVEY

As mentioned previously, the Alabama Gulf Coast is comprised of two large counties, Baldwin and Mobile. Although these counties are largely contiguous, they are quite different along many demographic factors. According to the 2010 census, there were just over 180,000 residents in Baldwin County, the median household income was about $50,000, and 13% of residents lived below the poverty line. There was little racial diversity, as 87% of residents were white and only 10% were African American. By contrast, there were just fewer than 413,000 residents in Mobile County, the median household income was $43,000, and 20% of residents lived below the poverty line. Residents were 60% white and 35% African American.

Both counties, however, are home to a massive tourism industry. As of the 2007 economic census, Baldwin County employed about 20,000 individuals and Mobile County employed nearly 35,000. As noted previously, estimates of the oil spill's economic impact on coastal Alabama range from several hundred million to $3.3 billion, and the majority of losses were in the tourism sector. With such significant impacts, there was a clear need to provide reliable baseline data on the effects of the *Deepwater Horizon* spill on coastal Alabama. At the time of our study, the only scientific survey of oil spill victims focused on residents in Louisiana and Mississippi. As such, our research filled a gap in knowledge and provided reliable information about the effects of the spill on the lives of Alabama residents. The primary goals of our pilot project were to understand the perceived causes, solutions, and long-term impacts of the oil spill.

To achieve those study goals, we partnered with the Survey Research Unit (SRU) at the University of Alabama–Birmingham (UAB) to conduct the first representative survey of Mobile and Baldwin County residents. The SRU operates within the Center for the Study of Community Health in UAB's School of Public Health, and has extensive experience in conducting scientific telephone surveys. It operates a 40-station call center with trained interviewers and the computer-assisted telephone interviewing (CATI) system at each station. SRU staff purchased access to a database of approximately 4,500 numbers from active landline and cellular telephone prefixes in the two Alabama counties under study. Once those databases were loaded into the CATI system, SRU interviewers used random digit dialing (RDD) to solicit participants.

Given the total population of nearly 600,000 residents in the two affected counties, we determined that approximately 400 completed surveys were needed to ensure a confidence interval of ±5%. The survey was administered by SRU interviewers over a 3-month period from February to April 2011. The overall response rate was 34%, and the completion rate was 26%. The total number of completed surveys was 402. Key topics from the survey instrument included (1) perceived economic effects of the spill on respondents, (2) perceptions as to why the spill occurred,

(3) perceptions of response effectiveness from local, state, and national authorities, (4) perceptions of the safety of continued offshore drilling and exploration, and (5) perceptions about prospects for recovery. Table 6.1 contains basic descriptive statistics for respondents in our sample.

Linking back to our conceptual framework of framing and blaming, we now present the results from the survey. Table 6.2 shows that just over 40% of respondents in our sample viewed the spill as an isolated incident, while almost 60% viewed it as part of a broader problem. Table 6.3

Table 6.1 Sample Demographics

Variable	Percentage
Gender	
Male	30.8%
Female	69.2%
Race	
American Indian	2.1%
Asian	0.5%
Black	21.8%
White	73.3%
Other	2.3%
Age (mean)	60.26
Education	
Less than high school	9.77%
High school (or GED)	25.06%
Some college	27.57%
4-year college degree	19.3%
Some graduate school	5.51%
Advanced degree	12.78%
Political views	
Very liberal or liberal	16.75%
Moderate	28.76%
Very conservative or conservative	54.84%

Table 6.2 Event Framing of Oil Spill

Q: Do you think that the BP oil spill was an isolated incident or an indicator of a broader problem with offshore drilling?	
Isolated incident	40.21%
Broader problem	59.79%

Table 6.3 Accountability for Oil Spill

BP Responsibility			**Federal Responsibility**		
	Frequency	**Percentage**		**Frequency**	**Percentage**
BP not at all responsible	5	1.26%	Federal government not at all responsible	75	19.43%
BP partially responsible	53	13.38%	Federal government partially responsible	208	53.89%
BP mostly responsible	117	29.55%	Federal government mostly responsible	58	15.03%
BP totally responsible	221	55.81%	Federal government totally responsible	45	11.66%

Cross-Tabulation

	Federal government not at all responsible	**Federal government partially responsible**	**Federal government mostly responsible**	**Federal government totally responsible**	**Total**
BP not at all responsible	1 0.26%	0 0.00%	1 0.26%	2 0.52%	4 1.04%
BP partially responsible	12 0.03%	30 7.77%	8 2.07%	2 0.52%	52 13.47%
BP mostly responsible	7 1.81%	79 20.47%	23 5.96%	7 1.81%	116 30.05%
BP totally responsible	55 14.25%	99 25.65%	26 6.74%	34 8.81%	214 55.44%
Total	75 19.43%	208 53.89%	58 15.03%	45 11.66%	386 100.00%

contains several tabulations of ascriptions of responsibility for the spill. The top left panel contains results for BP. The majority of respondents in our sample (55.81%) indicated that BP was totally responsible for the spill. On the other end of the spectrum, only 1.26% felt that BP was not at all responsible. The upper right panel shows the results for the federal government. Here, only 11.66% believed that the federal government was totally responsible, but a majority (53.89%) viewed the federal government as partially responsible. The bottom panel of Table 6.3 shows a cross-tabulation of the two variables. Most of the respondents believed that BP was mostly or totally responsible *and* the federal government was partially responsible (20.47% + 25.65% = 46.12%).

The analysis in the previous section showed how respondents in our sample viewed the event, as well as their attributions of acceptability. The final two elements in our discussion are risk perceptions and policy preference. Table 6.4 shows respondents' perceptions of future oil spill risk. The majority (52.9%) viewed future spills as a major threat, but more than 25% thought they were only a minimal threat.

Finally, we examine respondents' policy preferences. In our survey we asked respondents to make a choice using the following question: "Now I'm going to ask you about three policy alternatives. Please think carefully about how you feel about each one and tell me which you would most support." We then listed three policy alternatives that went from support of the status quo ("continuing offshore drilling, but with *no* new government oversight and regulation of the offshore drilling industry") to some new regulation ("continuing offshore drilling, but with *more* government oversight and regulation of the offshore drilling industry") to a total ban on new drilling ("banning all new offshore drilling"). Table 6.5 shows the distribution of policy preferences in our sample. Just over 22% supported the status quo (no new regulation). The vast majority supported some new type of regulation with continued drilling (67.62%). Only 10.18% supported a ban on new offshore drilling.

Table 6.4 Future Oil Spill Risk Perceptions

Q: Do you consider future oil spills to be a major threat, moderate threat, or minimal threat?	
Minimal threat	25.44%
Moderate threat	21.66%
Major threat	52.9%

Table 6.5 Policy Preferences

Q: Now I'm going to ask you about three policy alternatives. Please think carefully about how you feel about each one and tell me which you would most support.	
Continuing offshore drilling, but with *no* new government oversight and regulation of the offshore drilling industry	22.19%
Continuing offshore drilling, but with *more* government oversight and regulation of the offshore drilling industry	67.62%
Determinants of Risk Perception (Ordered Logit)	10.18%

AN INTERDEPENDENT MODEL USING PATH ANALYSIS

We now employ a path analysis to study the associations among our variables. The full path analysis is contained in Figure 6.1. We divided the path into three major stages to help fix ideas. In Stage 1, we examine the relationship between event framing and accountability. The results show that citizens who perceived the BP spill as an isolated incident viewed the federal government as less responsible (coefficient = –0.806). This lends support to our proposition that federal accountability is greater when systematic risks are perceived. That is, the federal government is held less responsible when disastrous events are viewed as isolated, but it is held more responsible when disasters are thought to result from systematic risks that should be regulated. Our results show no statistically significant relationship between

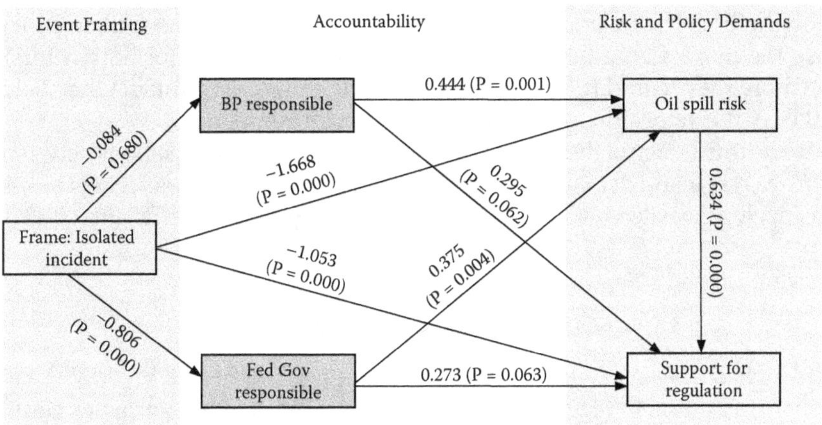

Figure 6.1 Path analysis.

event framing and BP's accountability, which suggests that while event framing may affect more distant accountability relationships, when there is a primary culprit, event framing makes little difference.

Event framing also has an impact on risk perception and policy demands. Narrow event framing was associated with less perceived risk of future oil spills (coefficient = –1.668). This also is in accord with our expectations. As stated earlier, the degree to which an event is viewed as isolated or part of a broader problem bears directly on levels of perceived risk. An isolated incident poses little future risk, but an event that is the manifestation of a broader problem is more likely to recur. There is also a relationship between event framing and policy demands. The negative relationship in the path demonstrates that narrow event framing is associated with less support for regulation (coefficient = –1.053).

The second stage of our model is accountability. Just as event framing has an impact on risk perception and policy demands, accountability also plays a role. The question of who is responsible for an incident is a special kind of frame that establishes the heroes and villains. We begin our analysis by looking at BP's accountability. We find that more ascribed accountability to BP is associated with greater perceived threat from future oil spills (coefficient = 0.444) and support for stricter regulations (coefficient = 0.295). Federal accountability is also positively associated with greater perceived threats from future oil spills (coefficient = 0.375) and more demand for regulation (coefficient = 0.273). Taken together, these results suggest that ascriptions of accountability have less of a differential effect on risk perceptions and policy demands. Regardless of whether people in our sample thought BP or the federal government was responsible for the spill, in either case the more accountability they ascribed to either actor (or both actors), the more they feared future oil spills. Also, the more accountable they held either of the actors, the greater their demand for regulation.

The third stage of the model involves examining the associations between perceived risk and demand for regulation. We find that more perceived risk is positively associated with demands for stricter regulation (coefficient = 0.634).

MULTIVARIATE ANALYSIS

The path analysis is a useful tool for examining the simple bivariate relationships among the factors in our model, but many other variables could be at work. To test the robustness of our findings, we present the results of multivariate analyses that contain a large number of additional control variables. We include controls for education (highest grade completed), political conservatism (a 5-point scale), age, gender, and race. Table 6.6 contains the results for risk perception.

Because the risk variable is a three-level ordinal variable, we use ordered logit in our analysis. The first column contains the baseline results, which are identical to those contained in the path analysis. Narrow event framing is negatively associated with perceived oil spill risks in the future (coefficient = 1.668). Greater accountability for BP or the federal government is associated with higher perceived risks. The second column shows the full model with all of the control variables added. Even with the inclusion of the additional variables, the basic findings from the baseline results are unchanged. The three baseline factors are still statistically significant, with coefficients very near to their original values. This suggests that the

Table 6.6 Determinants of Perceived Risk (Ordered Logit)

	(1) Baseline	**(2) With Controls**
Isolated incident	–1.668***	–1.687***
	(0.219)	(0.246)
BP responsible	0.444***	0.407***
	(0.135)	(0.148)
Federal government responsible	0.375***	0.382***
	(0.131)	(0.145)
Education		–0.130
		(0.0792)
Political conservatism		–0.199
		(0.133)
Age		–0.0234***
		(0.00790)
Female		–0.103
		(0.236)
Nonwhite		0.112
		(0.266)
N	383	349
Pseudo-R^2	0.116	0.139

Note: Dependent variable is perceived threat from future oil spills. Positive coefficients increase perceived threat. Standard errors in parentheses.
$^*p < .10$, $^{**}p < .05$, $^{***}p < .01$.

underlying relationships established in the path analysis are robust to differences in model specification. Among the control factors, we find that age is negatively associated with perceived risk. The other control factors are not statistically significant.

Table 6.7 shows the results for regulatory support. The regulation variable was also a three-level ordinal variable, so ordered logit is used in this analysis as well. The first column contains the baseline results. As we noted earlier, perceived risk is positively associated with demands for stricter regulation, as are ascriptions of accountability to either BP or the federal government. In contrast, narrow event framing is associated with less demand for regulation, as evidenced by the negative coefficient on the

Table 6.7 Determinants of Support for Regulation (Ordered Logit)

	(1) Baseline	(2) With Controls
Oil spill risk	0.634***	0.449***
	(0.165)	(0.173)
Isolated incident	–1.053***	–1.100***
	(0.256)	(0.274)
BP responsible	0.295*	0.345**
	(0.158)	(0.165)
Federal government responsible	0.273*	0.299*
	(0.147)	(0.156)
Education		–0.0623
		(0.0908)
Political conservatism		–0.575***
		(0.160)
Age		–0.0224***
		(0.00869)
Female		0.457*
		(0.259)
Nonwhite		0.0786
		(0.284)
N	371	340
Pseudo-R^2	0.118	0.175

Note: Dependent variable is support for stricter regulation. Positive coefficients increase support for regulation. Standard errors in parentheses.
*$p < .10$, **$p < .05$, ***$p < .01$.

event framing variable. The next column shows the results with controls added. As with our risk model, the baseline results are largely unchanged with the addition of the controls. We do find some interesting relationships between policy demands and the controls, however. Unsurprisingly, political conservatism has a negative association with support for regulation. We also find a negative association between age and support for regulation. Women showed more support for regulation than their male counterparts, as evidenced by the positive coefficient.

CONCLUSION

In this chapter we used the social amplification of risk framework to examine the relationships among event framing, accountability, risk, and policy preference. Based upon our empirical analysis, there were several noteworthy findings. First, event framing matters; we found that event framing had an impact on whether actors with regulatory responsibility were held accountable. If an event was viewed narrowly as an isolated instance, there was less accountability for regulators. If it was viewed as part of a broader problem, regulators were held accountable. This finding coalesces with the social amplification of risk framework. Risk amplification can occur through the media, making events seem bigger and broader than their framing. To the extent that the frame is broadened enough, regulatory actors, like the federal government, are held responsible. Second, event framing also had a direct impact on the perceived risk of recurrence. Isolated incidents are viewed as random results of bad luck, but events that are portrayed as part of larger and systematic causes are more likely to recur.

Accountability also matters. We found that regardless of whether BP or the federal government is held responsible for the spill, higher degrees of accountability are associated with higher perceived risk of future spills and more demand for regulation. This is an interesting contrast to our findings on event framing. One might think that if a person held BP accountable for the spill and not the federal government, he or she might not favor regulation as a solution, but our empirical model did not suggest this. This may also have to do with the nonexclusive nature of our survey design. We allowed respondents to independently rate the accountabilities of BP and the federal government. Recalling the pattern we saw in survey responses (that most thought BP had a great deal of responsibility for the spill and the federal government was partially responsible), this could explain why ascribed accountability to either actor was associated with more demand for regulation.

Finally, the logics of event framing, accountability, and risk come to bear upon the relationship between framing and policy demands. Broad problems require a broad response, and broader responses are usually in the purview of regulatory actors. Therefore, a broad event framing will likely result in demands for increased regulation, whereas a narrow event frame might lead to localized demands for liability in the courts, but not an overarching approach.

ACKNOWLEDGMENTS

Funding for this study was provided by the University of Alabama at Birmingham via a competitive Pilot Projects Grant awarded in October 2010. The authors thank Paul Wolff and personnel at UAB's Survey Research Unit for assistance in data collection.

REFERENCES

Addy, Samuel, and Ahmad Ijaz. *Some Preliminary Macroeconomic Impacts of the 2010 Deepwater Horizon Oil Gusher on Alabama*. Technical Report. Tuscaloosa, AL: University of Alabama, Center for Business and Economic Research, 2010. http://cber.cba.ua.edu/Prel_MacroEcon_Impacts_on_AL_of_2010_BP_Oil_Spill.pdf.

Beamish, Thomas D. Waiting for crisis: Regulatory inaction and ineptitude and the Guadalupe dunes oil spill. *Social Problems* 49, no. 2 (2002): 150–77.

Bozeman, Barry. The 2010 BP Gulf of Mexico oil spill: Implications for theory of organizational disaster. *Technology in Society* 33, no. 3–4 (2011): 244–62.

Diel, Stan. $573 million in BP oil spill claims paid in Alabama, but some who qualify haven't filed, AG says. 2013. http://www.al.com/business/index.ssf/2013/06/573_million_in_bp_oil_spill_cl.html (accessed September 13, 2014).

Douglas, Mary. Risk and blame. In *Risk and Blame: Essays in Cultural Theory*, ed. Mary Douglas, 3–21. New York: Routledge, 1992.

Freudenberg, William R. Institutional failure and the organizational amplification of risks: The need for a closer look. In *The Social Amplification of Risk*, ed. Nick Pidgeon, Roger E. Kasperson, and Paul Slovic, 102–120. Vol. 1. Cambridge: Cambridge University Press, 2003.

Freudenberg, William R., and Robert Gramling. *Blowout in the Gulf: The BP Oil Disaster and the Future of Energy in America*. Cambridge, MA: MIT Press, 2012.

Frewer, Lynn J. Trust, transparency, and social context: Implications for social amplification of risk. In *The Social Amplification of Risk*, ed. Nick Pidgeon, Roger E. Kasperson, and Paul Slovic, 123–137. Vol. 1. Cambridge: Cambridge University Press, 2003.

Gochfeld, Michael, and Joanna Burger. Ecological risk, risk perception, and harm: Lessons from the Arthur Kill. In *Before and After an Oil Spill: The Arthur Kill*, ed. Joanna Burger, 265–279. New Brunswick, NJ: Rutgers University Press, 1994.

Hoffman, Andrew J., and P. Devereaux Jennings. The BP oil spill as a cultural anomaly? Institutional context, conflict, and change. *Journal of Management Inquiry* 20, no. 2 (2011): 100–12.

Kasperson, Jeanne X., Roger E. Kasperson, Nick Pidgeon, and Paul Slovic. The social amplification of risk: Assessing fifteen years of research and theory. In *The Social Amplification of Risk*, ed. Nick Pidgeon, Roger E. Kasperson, and Paul Slovic, 13–46. Cambridge: Cambridge University Press, 2003.

Kasperson, Robert E., Ortwin Renn, Paul Slovic, Halina S. Brown, Jacque Emel, Robert Goble, Jeanne X. Kasperson, and Samuel Ratick. The social amplification of risk: A conceptual framework. In *Social Contours of Risk: Volume I: Publics, Risk Communication and the Social Amplification of Risk*, ed. Jeanne X. Kasperson and Roger E. Kasperson, 99–114. London: Earthscan, 2005.

Kingdon, John. *Agendas, Alternatives, and Public Policies*. 2nd ed. Reading, PA: Addison-Wesley, 1995.

National Commission on the BP Deepwater Horizon Oil Spill and Offshore Drilling. Deep Water: The Gulf Oil disaster and the future of offshore drilling. Report to the president. 2011.

Oliver, Mike. Alabama could reap billions due to judge's ruling today in BP oil spill, attorney general says. http://www.al.com/news/index.ssf/2014/09/alabama_could_reap_billions_af.html (accessed September 13, 2014).

Palomo, Juan. API response to commission report: 'We've made progress to improve safety.' Press release, January 11, 2011. http://http://www.api.org/news-and-media/news/newsitems/2011/jan-2011/api-response-to-commission-report (accessed November 9, 2014).

Pidgeon, Nick. The limits to safety? Culture, politics, learning and man-made disasters. *Journal of Contingencies and Crisis Management* 5, no. 3 (1997): 1–14.

Pidgeon, Nick, and M. O'Leary. Man-made disasters: Why technology and organizations (sometimes) fail. *Safety Science* 34, no. 1–3 (2000): 16–30.

Reuters. Timeline—Gulf of Mexico oil spill. July 23, 2010. http://www.reuters.com/article/2012/11/15/us-bp-spill-fine-idUSBRE8AE0WC20121115 (accessed September 13, 2014)..

Weber, Harry R. Blown-out BP well finally killed at bottom of Gulf. *Boston Globe*, September 19, 2010. http://www.boston.com/news/nation/articles/2010/09/19/blown_out_bp_well_finally_killed_at_bottom_of_gulf (accessed September 13, 2014).

7

How Government's Actions after a Disaster Affect Long-Term Civic Engagement

Shifting Opportunity and Motivation for Civic Participation in Christchurch, New Zealand

Stephanie Hawke, Jillian Girard, and Jane Carr

Contents

INTRODUCTION

In 2011, a yet undiscovered fault line shifted abruptly 7 kilometers below the city of Christchurch, the second largest city in New Zealand, creating an earthquake with a 6.3 magnitude. Due to the shallowness of the earthquake, nearly three-quarters of the city's housing stock was destroyed. The earthquake is embedded in the collective memory of Christchurch residents, but another story resulting from this disaster remains largely untold. Local and national government responses to the disaster, which shifted throughout the response and rebuilding periods, yielded a significant impact on civil society and civic engagement (Bennett, 2011). While we often understand civic engagement as a function of factors such as citizens' financial, educational, and time resources (Foster-Bey, 2008), we argue here for more inclusion into this equation of governments' ability to provide meaningful opportunities for input following a disaster. Studying the disaster context serves a dual purpose: it yields deeper knowledge about resilience, which can be applied by practitioners, but disasters also create a unique political landscape in which rapid changes in governance strategies and citizen resources allow for a close examination of how these factors interplay.

Using the experience of the 2011 Christchurch earthquake as a case study, we explore the change in civic engagement over time and propose a framework for understanding how to maximize engagement in the rebuilding period. This study is not focused on the time period immediately following the earthquake, in which civic relationships are forged out of unifying adverse circumstances, but rather looks at the long-term impact such an event can have on civic engagement and active political participation in democratic governance. We argue that civic engagement should be a core feature of rebuilding a community in a postdisaster context due to the long-term effects choices made in rebuilding have for the community. Further, we hypothesize that when opportunities for civic engagement are limited following a disaster, the effect has long-term negative consequences on the overall civic engagement of a community. A disaster offers many communities an opportunity to reinvent their historical structures in light of new experiences; however, the consequences may not always align with their vision.

THE CHRISTCHURCH EARTHQUAKES

A series of earthquakes rocked the Canterbury region of New Zealand between 2010 and 2011. This seismically tumultuous period was punctuated by two major earthquakes: the first in September 2010 and the second in February 2011. Though the former was rated as stronger on the Richter scale, the latter was by far the most damaging. This section of the chapter provides a brief overview of the events, damage, and costs associated with each earthquake.

On September 4, 2010, the Canterbury region experienced a 7.1 magnitude earthquake. The event was centered 37 kilometers outside of Christchurch, on a fault line that was previously undiscovered (Adetunji, 2010). Despite the violence of the quake, the damage was limited: two people were injured and none were killed. The New Zealand Army was called upon to maintain regional peace, but this, as it turned out, was unnecessary. Residents were resilient to the disaster. They called upon social networks, community ties, and local government (Christchurch City Council [CCC]) for support and security. The very powerful earthquake could have been devastating—instead, it had little lasting impact on the region.

The aftershock in February 2011 presents a different story. Another previously undiscovered fault line, which ran under the center of town, slipped. This resulted in a 6.3 magnitude earthquake. The epicenter of the quake was less than a mile from the city center and only 7 kilometers deep. Some buildings collapsed immediately; the CTV building, home to a foreign language school and a local news studio, was destroyed and 115 people were killed instantly (Taylor, 2011). Other buildings listed precariously, a full public bus was crushed by falling brick and beams (Turner,

2012), and the façade and spire of the iconic Christchurch cathedral toppled (Quinn and Tran, 2011).

In all, 185 people died and 150,000 homes were damaged; 10,000 buildings were eventually demolished (Parker and Steenkamp, 2012). The city center was cordoned off for months. Huge piles of liquefaction (silt that bubbles up from the soil) erupted from the cracks in the pavement, leaving the city a strange, sandy desert. More than 3% of homes were eventually declared to be in a red zone, meaning that any further use or building on the property was prohibited (Chang et al., 2014). Local response was not enough; established stores of resilience were not going to save the region this time.

Christchurch is the second largest city in New Zealand; the disaster affected the national gross domestic product (GDP), unemployment, manufacturing, and tourism (Cohen, 2014; Ministry of Business, Innovation and Employment, 2013). It was clear that national support was required. The National Crisis Management Centre (NCMC) coordinated with the CCC for the emergency response. The rescue efforts—which included search and rescue teams from countries as far as South Korea, Japan, the United States, and Mexico—were managed by NCMC with the help of the CCC. Distribution of emergency provisions such as chemical toilets and water were run through the CCC. Tents, blankets, and food were made available to everyone; camping was encouraged in Christchurch's central park. In all, the immediate response to the February 2011 quake was exemplary in terms of vertical governance cooperation. The local government of Christchurch, with national support from the NCMC, navigated the response very well. The rebuilding period, however, did not display the same effective cooperation.

Response and Rebuilding

In the months immediately following the earthquake, the CCC headed up rebuilding efforts. Two months after the immediate response, the council created Share an Idea: a program that invited all residents of Christchurch to help imagine the city of the future. People were asked to write in and propose ideas and goals for what Christchurch could become. Respondents could visit a local site in person, mail in their ideas, or submit on a hosted website. The project was open for 6 weeks and collected 100,000 ideas. More than 21% of city residents participated (NV Interactive, n.d.). "Share an Idea has been lauded throughout the world," receiving "numerous international prizes for both its inventiveness and its openness in responding to a crisis situation" (Carlton, 2013, p. 8). The ideas were analyzed and distinct trends emerged, such as increased bikeways, mixed-use developments, and a concentrated, walkable downtown. City planners took the results of Share an Idea and developed the central city plan—a guide to rebuilding Christchurch (Carlton, 2013, p. 9). Share an Idea created a discursive planning process, and with the implementation of these ideas in the central city plan. Share an Idea seemed to be the pinnacle of "empowered deliberative democracy" (Fung and Wright, 2001, p. 7).

The Share an Idea story is not one of optimal civic engagement, however. What started out so auspiciously met a lackluster end. As noted above, few doubted that national support was necessary for Christchurch to fully recover. In 2011, the national parliament passed the Canterbury Earthquake Recovery Act (CER Act). One of the primary elements of the bill was the creation of the Canterbury Earthquake Recovery Authority (CERA), an administrative agency charged with rebuilding the Canterbury and Christchurch regions. The organization of the agency was stipulated by the CER Act—agency leaders would be appointed by the prime minister (Brookie, 2012). There was no local accountability built into the CERA. Leaders were accountable to the prime minister; therefore, the people of Christchurch had no recourse to contest decisions (Carlton, 2013, p. 3). CERA was charged with two main tasks. The first was to lead a "time-bound rebuild"—to make Christchurch and the surrounding region functional as soon as possible. The other mission was community outreach. As Carlton (2013) notes, these goals were conflicting. Community outreach takes time; CERA was charged with getting Christchurch back to normal as soon as possible.

From its inception, the central city plan was supposed to be utilized as the guiding document for rebuilding Christchurch. CERA's intervention, however, called this into question. The agency quickly established the Christchurch Central Development Unit (CCDU) with a mission to *create* rebuilding plans for the central city. It seemed increasingly unlikely that the CERA would merely implement the plan as devised by Christchurch residents. Christchurch leaders, too, were concerned about the replacement of the locally devised plan. Mayor Bob Parker directly assuaged trepidation with the CERA takeover, reassuring Christchurch residents that the document—and related research—would not be discarded. Leaders in Christchurch believed that they would work alongside CCDU in creating a new plan, informed by the results of Share an Idea (Gates et al., 2012). However, as Edwards (2009) theorizes, interjurisdictional government cooperation is difficult, particularly after a natural disaster. The intergovernmental tension of a cross-boundary event proved too much. Cooperation was quickly replaced by unilateral decision making by the CCDU (Carlton, 2013). The central city plan was wholly discarded. The CCDU version was objectively and markedly different. Funds were earmarked for projects—such as a new rugby field—that had not been identified as priorities by Christchurch residents through Share an Idea (Chang et al., 2014).

Certainly, decision makers in Christchurch faced considerable challenges during this period. Yet there was an observable change in the way that decisions were made. The CCC started with a community-focused, participatory planning approach. There was then a stark "shift in ownership of the central city rebuild plans from the City Council to the CCDU" (Carlton, 2013, p. 10). The CCDU planning was characterized by less community engagement, and a shift in priorities that mirrored national, as opposed to local, goals.

The Christchurch earthquake provides us with an opportunity to study civic engagement prior to the disaster, during the response period, and during the long-term rebuilding period. While some attention has been given to the response of citizens immediately following a disaster, there has been relatively little scholarly work done regarding the long-term impact of disaster on civic engagement. We hope to add to this literature by presenting a framework for understanding the potential long-term impacts of a disaster on civic engagement, followed by a review of the Christchurch case example in light of this framework. We suggest that given the major social, political, economic, and other impacts a disaster has on a community, the potential for long-term community changes is likely, and perhaps inevitable. We then call for greater attention to be paid to the disaster rebuilding efforts related to civic engagement. During the postdisaster rebuilding process community norms may be reformed. Thus, institutionalizing a culture of participation, trust, and sense of community can be just as important as the physical act of rebuilding.

First, we must establish what we mean by civic engagement, opportunities for participation, and civic motivation. Following a discussion of these concepts, we will describe our framework and demonstrate how we believe these concepts interact at the community level. Finally, we will revisit the case study described above to provide a more tangible example of the framework.

MOTIVATION AND OPPORTUNITIES FOR CIVIC ENGAGEMENT IN A POSTDISASTER CONTEXT: KEY LITERATURE AND CONCEPTS

Civic Engagement

Although varying definitions of civic engagement have been proposed, there is strong consensus that higher levels of participation in the form of activities such as voting, volunteering, participating in community organizations, donating money, and contributing to political advocacy lead to more effective governing institutions and more representative governing decisions. In their review, Adler and Goggin (2005) define *civic engagement* as "how an active citizen participates in the life of a community in order to improve conditions for others or to help shape the community's future" (p. 241). Two elements of this definition are particularly important here: the active citizen and the intention of civic engagement is to help shape a community's future. Civic engagement involves citizens actively participating in public life, and there is a critical distinction between a desire to participate and actively participating in public activities. Action by itself does not represent civic engagement. Rather, the role of individuals and organizations working toward a perceived common good elevates mere action to civic engagement.

The role and impact of civic engagement has been thoroughly studied by numerous scholars. Putnam (1993) describes the relationship between civic engagement and effective government and argues that a higher degree of civic engagement leads to more effective government operations due to social capital development. However, additional scholars (Levi, 1996; Theiss-Morse and Hibbing, 2005) added to the work of Putnam and argued that the voluntary associations, which Putnam identifies as his social capital–building mechanisms, are insufficient to produce the strong political culture he describes.

State institutions can serve as incubators or restrictions on civic engagement (Schofer and Fourcde-Gourinchas, 2001; Skocpol et al., 2000). Civil society does not exist in a vacuum and the state is a powerful partner. Skocpol (1997) contends, "From early on in America's national history, the structure and activities of the federal government, along with translocal and competitive forms of popular political mobilization, created an 'opportunity structure' that nourished, encouraged, and rewarded voluntary associations" (p. 472). It is not exclusively the voluntary associations that foster civic engagement, but also the structure and functioning of the government in which those associations are embedded.

Building on this concept, Ekman and Amna (2012) suggest two broad categories of civic engagement: civil participation, similar to Putnam's definition earlier, and political participation. Further, Adler and Goggin (2005) posit that not all activities are equal in intensity. Voting, for example, while an important form of political action, does not have the same impact as holding public office. Thus, while a single term, *civic engagement*, is used to describe this multitude of actions with varying intensity, it is important to note the rich variety of actions encompassed in this phrase. Importantly, we are conceptualizing civic engagement here in the broad sense of the term, which is inclusive of both the political and social aspects.

Civic Engagement in a Disaster Context

Disasters affect a community's social, political, economic, and psychological resources, and because of this, they affect civic engagement (Aldrich, 2011). People's time and ability to participate are drastically affected by a disaster (Foster-Bey, 2008). As a result, the level of civic engagement may vary dramatically between the periods prior to and immediately following (response period), and ultimately, in the rebuilding period that follows the response (Simo and Bies, 2007). Significant scholarly attention has been given to examining civic engagement in the context of disasters, specifically during the response period (Aldrich, 2010; Birkmann et al., 2008; Shaw, 2004). A robust body of literature on resilience revealed numerous ways in which social capital and civic engagement interact with other factors to shape, and be shaped by, response to disasters. The great majority of this scholarly work examines how social capital levels predisaster enable recovery. Neighborhood networks allow for community members to support one another, serve those in need directly, and also spread awareness and vital news in the wake of disaster (Aldrich, 2010). Trust in community associations and governmental institutions also allows government and nongovernmental organization (NGO) response efforts to be more readily received, and increases the likelihood of compliance with disaster management requests.

Aldrich (2010) provides one example of this strand of scholarship, arguing that material resources are far from the sole determinants of resilience. He states that "governmental and NGO response to disasters has been premised on the idea that moving more money, supplies, and experts into affected areas will result in a faster recovery" (p. 2). Instead, Aldrich (2012) and others (Berke and Campanella, 2006; Chamlee-Wright and Storr, 2011; Cox and Perry, 2011; Shaw, 2004) propose that social capital is a critical and necessary component of community resilience in a postdisaster context. Additionally, Putnam (1993) argues that the ability to trust one another and work collaboratively builds community networks and fosters civic engagement. Applying Putnam's theory to the disaster context, we see how the informal networks that build social capital create the basis for collective action during the response and rebuilding periods. Thus, a community high in social capital is more resilient than a community with low social capital. While this avenue of scholarship is essential for improved disaster management practice, another question about civic engagement in the disaster context remains largely unstudied: How does a major disaster impact citizens' *long-term* willingness, desire, and ability to participate in public decisions?

Motivation for Political Participation

The interplay of the individual and the state, sometimes through organizations, is a critical component of democracy. Thus, the role of the individual is an important consideration here. By *motivation for participation* we specifically mean the collective investment, interest, and ability of individuals to participate in public life and governance. This includes interacting with community in formal and informal organizations (civic participation), in addition to participating in the democratic process (political participation). Though seemingly straightforward, the idea of political participation opportunity is somewhat nebulous in the field. Meyer (2004) highlights the lack of concurrence between scholars, writing that "conceptualizations of political opportunities vary greatly"; these differing conceptualizations necessarily lead to differing analyses (pp. 125–126). Our conception of *opportunity for participation* picks up necessary threads in the literature to construct an appropriate concept for our purposes here.

A handful of scholars posit that an individual's motivation for political participation is guided by an internal cost–benefit analysis. Leighly (1995) discusses how perceived potential benefits define the amount of effort individuals are willing to commit to a cause. This, in turn, defines the types and amount of desired participation opportunities. If citizens see a minimal benefit to participation, they will commit limited resources; if, on the other hand, they see the potential gain from their actions as very large or transformative, they will seek more opportunities to participate in government. This internal calculus (Klandermans, 2009; Stolle and Hooghe, 2005) is the bedrock to our theory. When citizens have resources and motivation to impact community decisions, and perceive that their efforts can result in tangible change, they desire more participation opportunities.

The second important theme in the field is the idea of social capital. For some authors, a high level of social capital implies a high level of motivation to have meaningful political impact. Kirlin and Kirlin (2002) allude to Putnam's work when they identify three factors as determinants of demand: motivation, skill, and network connections akin to those described by Putnam. Network connections are essential drivers of both motivation (connections between the individual and associations) and opportunities (connections between central decision makers and community associations and organizations).

Though we may take for granted the desire of citizens to participate in a democracy, there are many factors that lead individuals toward disengagement from the public sphere. Perceptions of possible impact drive motivation; when citizens view their impact as naught, they are less likely to seek out participation opportunities. When citizens are isolated—socially or politically—they become less interested in solving communal problems. It is also important to note, as highlighted by Aldrich (2012) and Fiorina (1999), that strong social ties among some citizens can also have negative consequences for citizens who are less represented in the community's social capital network.

Opportunities for Participation

The availability of opportunities for civic participation is necessarily driven by political institutional factors, as Berry et al. (1993) highlight. They write that government administration, funding, communication, and city government structure all affect the number of participation opportunities available to the public. This echoes the thesis of Koliba et al. (2010), who argue that long-standing networks allow for the development of informal connections among nodes. These informal connections speed information dissemination, facilitate trust, and essentially bolster the formal process. Long-standing networks that connect the public with central decision makers are characterized by this network of informal connections.

UNDERSTANDING LONG-TERM CHANGES IN CIVIC ENGAGEMENT AFTER DISASTERS: INTRODUCING A NEW FRAMEWORK

The long-term impact of natural disasters upon civic engagement is not well understood. Certainly, this issue is difficult to address empirically given the multiple inputs to civic engagement in society, which are complicated further by disaster. However, by conceiving civic engagement as a product of shifting motivation and opportunities for political input, we can map out key ways in which a government's actions after disasters will affect civic engagement over time. Our framework helps explain changes in civic engagement observed in Christchurch, and is applicable to other developed democracies. It explains some of the unique challenges for governmental leaders after a disaster, pointing to how and why civic engagement can suffer in the long run from what is otherwise an effective disaster response. We argue that these lasting impacts have been underestimated and require more scholarly attention.

The foremost role of government after a disaster is to save lives. The benefits of acting quickly far outweigh the costs to participatory governance. Yet how can we understand the transition process by which administrators transition from a mode of rapid, more centralized decision making to a mode of slower, but more inclusive decision making? The benefits of this more inclusive mode include not only increased legitimacy and stability, but also the long-term strengthening of democratic culture and satisfaction of citizens.

Impact of Disaster on Civic Engagement

The initial availability of opportunities for participation may dwindle where response efforts are more centralized and rapid. In these instances, the rapid response limits the ability of government to include the full range of input from the public. Conversely, where government is more inclined to go through the process of input, the progress may be slower but more inclusive. In many, although not all, disaster response efforts, the government's first response is more centralized (Harrald, 2006). However, once the period of response ends and the rebuilding period begins, leaders may discern that the opportunities for broader participation have reemerged, and will actively build opportunities to participate in decision making. This discernment, however, is a difficult step. Leaders have to balance time and effort expended in the face of a disaster against civic opportunities to participate in the process. In the rebuilding period following the Christchurch earthquake, for example, an inclusive process was implemented first, followed by a more centralized response from the national government. So, while there was an initial increase in the availability of opportunities for participation, there was a decline when the responsibility for disaster relief changed hands from the local to the national level.

Simultaneous to a shift in availability of opportunities for participation—which could be either more or less extensive than before a disaster—a shift in motivation for involvement may occur (Ursano et al., 2007). For some, the response to a disaster is clearly to manage personal or family crises. For many, there will be a period of support followed by feelings of loneliness or mistrust (Ursano et al., 2007). As with opportunities for participation, the motivation to participate will vary. While the individual's internal calculus will inform the decision whether to engage in civic engagement activities, action will also depend on the available opportunities for participation. Again, it is important to note the focus of this chapter is not the time period immediately following the disaster, but the rebuilding period that follows.

A disaster is a moment in time that can be identified as a disruption of normal routines (Aldrich, 2012). It is at this point that patterns of civic engagement may change quite dramatically within society because both access to opportunities and each individual's inclination to participate will be evaluated anew. Availability and interest either increase or decrease in the period following a disaster. We suggest that both opportunity and motivation must be high in order for civic engagement to occur at the highest potential level. Table 7.1 demonstrates the hypothesized civic engagement levels for each of the four quadrants (high and low motivation by high and low opportunity). If either or both motivation for participation and opportunity for participation are low, the resulting

Table 7.1 Matrix of Opportunity and Motivation

	High Availability of Opportunities for Participation	Low Availability of Opportunities for Participation
Highly motivated individuals	High civic engagement	Low civic engagement
Low motivation among individuals	Low civic engagement	Low civic engagement

level of civic engagement will also be low. However, if both are high, civic engagement will be high as well. We suggest that actual participation should be measured using the robust understanding of civic engagement suggested by Adler and Goggin (2005), which includes both civic and political participation. By motivation we mean the individual's decision-making process, which accounts for preference as well as socioeconomic factors. Finally, in regard to the opportunity for participation, we turn to Berry et al. (1993) and suggest the concept is operationalized through examining city resources, programs, communication mechanisms, and administration.

CHRISTCHURCH: AN EXAMPLE OF THE FRAMEWORK

As often occurs after a disaster of this magnitude, the governmental response prioritized efficiency over participatory decision making in the months following the 2011 Christchurch earthquakes (Harrald, 2006). Leaders see opportunities for broad engagement in the longer-term rebuilding process, yet nonetheless must move forward at the outset without inclusion. The challenge for political leaders is to discern when and how the public can once again be broadly included. According to our framework, if the difficult challenge of rebuilding participation opportunities is not met, citizens will begin to feel distanced and then dissatisfied with the decisions being made, resulting in a decrease in civic engagement. In the case of Christchurch, Share an Idea provided abundant opportunities to engage, and its demise is linked to a negative impact on citizen approval of decision making.

As access to and interest in participating shift after a disaster, levels of concrete civic engagement actions respond. In the initial period of altruism and organizing, citizens will join more associations, volunteer, and be more likely to partake in existing opportunities for participation, such as voting. In Christchurch, the Share an Idea program was timed to capture this enthusiasm for participation. It provided a unique opportunity for a more open-ended, dynamic, and creative type of feedback to emerge from citizens, in comparison to citizens casting ballots. As noted previously, Share an Idea received a tremendous response with over 100,000 replies. This not only represents an accessible opportunity for civic engagement, but also is noteworthy evidence of civic engagement (the action taken by individuals to benefit the community) following the 2011 earthquake.

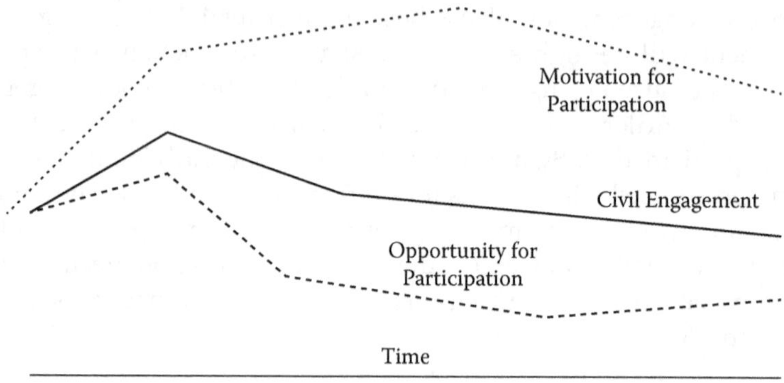

Figure 7.1 Impact of motivation and opportunity for participation on civic engagement.

However, as the access to opportunities subsided, the framework would suggest a decline in civic engagement (limited opportunity leading to limited civic engagement). A related effect is the decline in the population's trust in governing institutions. This results from a spreading awareness among the increasingly networked population that attempts to influence governing bodies are likely to be unsuccessful. As part of this awareness, participation in voluntary organizations and elections subsides below predisaster levels (Figure 7.1).

Changing Opportunity, Motivation, and Civic Engagement in Christchurch

The Christchurch, New Zealand, earthquakes of 2010 and 2011 were selected together as a case study, both to preliminarily assess the framework's applicability to a particular case and to contribute to an understanding of these particular disasters. Data about the opportunities and motivation for participation and civic engagement response following the Christchurch earthquakes were reasonably available. We drew upon the General Social Survey data (Statistics New Zealand, 2012) and the Quality of Life Survey,* which provides data from 2008, 2010, and 2012. To examine the pairwise comparisons, we performed a one-way analysis of variance (ANOVA). Additionally, we drew upon the CERA Wellbeing Survey 2014, which provides data from

* Raw data from 2008, 2010, and 2012 Quality of Life Survey administrations graciously provided by the Christchurch City Council.

Table 7.2 Indicator Variables Used to Examine the Relationship among Motivation to Participate, Opportunity to Participate, and Civic Engagement

	Motivation to Participate	Opportunity to Participate	Civic Engagement
Indicator	• Confidence in the government (CCC or CERA) • Understanding how decisions are made • Desiring more influence over local decisions	• Accessible entry point for citizen engagement • Meaningful opportunities for citizen engagement	• Voter turnout • Membership in organizations • Volunteerism

September 2012, April 2013, September 2013, and April 2014 and reports significant differences at the 95% confidence level for each year. We also drew on additional election data provided by the government of New Zealand and utilized McNemar's test to determine significant changes from year to year. Its unique response provided a particularly rich example of how opportunities and motivation interact to lead to surprising net outcomes in civic engagement. See Table 7.2 for the list of indicator variables.

Disaster Effects on Motivation to Participate

Individuals weigh not only the opportunities available for input, but also their own perception of government, which shapes their determination of whether participation is worth the investment. In this case analysis, three variables provide insight into individuals' motivation to participate:

1. Confidence in government entities (both the city council and CERA). If citizens do not have confidence in their government, their desire to participate will suffer. At the core of motivation is the belief that participation is worthwhile.
2. Understanding how decisions are made. If citizens do not understand how their government works, it will be more difficult for them to engage in its processes.
3. Whether an individual wants more input in city council decisions.

These three variables were measured using secondary data from a number of surveys. The survey questions pertaining to the city council come from the New Zealand Quality of Life Survey, with results from 2008, 2010, and

2012. The question pertaining to CERA, or confidence in CERA to make the best decisions for the area, is from the New Zealand Wellbeing Survey administered by CERA and is available for 2012, 2013, and 2014.

Following the first earthquake in 2010, there was a spike in Christchurch residents' confidence in government. The percent of respondents indicating agree or strongly agree increased from 23.9% in 2008 ($N = 486$, mean [M] of 2.60 on 5-point scale) to 53.5% in 2010 ($N = 496$, $M = 3.41$) ($p < .05$). After the second earthquake we expect that the confidence in city council would decline based on CERA's takeover of the earthquake management and the corresponding decline in city-level management. This was, in fact, exactly what occurred. There was a major decline in the percentage of people agreeing or strongly agreeing with the statement "I have confidence that the council makes decisions in the best interest of the city," dropping from 53.5% in 2010 ($N = 496$, $M = 3.41$) to 32.0% in 2012 ($N = 560$, $M = 2.70$) ($p < .05$). Similarly, for CERA, 41% of respondents indicated that they are confident (including "very confident" or "confident") that CERA is making the best decisions for the area in September 2012. However, by April 2013, that had dropped significantly to 35% of respondents. The confidence remained around April 2013 levels for both September 2013 (35%) and April 2014 (33%) (CERA Wellbeing Survey 2014). This trend suggests that individuals' motivation to participate may be affected by a declining trust in government.

Additionally, for the variable "I understand how my council makes decisions," the same pattern was observed. Between 2008 ($N = 479$, $M = 2.82$) and 2010 ($N = 496$, $M = 3.11$) there was an increase in people agreeing or strongly agreeing with the statement, from 34.1% to 37.7% ($p < .05$). Conversely, in 2012 ($N = 555$, $M = 2.64$), the agreement with understanding council decisions dropped to a low of 24.8% ($p < .05$). This indicates that people felt they understood how council made decisions less in 2012 than in either 2010 or 2008. We expect individual-level confusion to emerge as a result of the national government's takeover of planning and development. This confusion results in less motivation for participation because people simply do not know how to access the system.

Finally, in 2008, 58.2% of survey respondents reported that they would like more say in the council ($N = 482$, $M = 3.69$). We anticipate that the percentage of respondents who want more of a say in council will decline as representation increases. That is, as constituents have more confidence in their councilors, demand for a public voice will decline. In 2010, the percent of people agreeing with the statement "I would like to have more of a say in what the council does" dropped to 45.8% ($N = 496$, $M = 3.42$)

Table 7.3 Survey Responses Used to Assess ***Motivation to Participate***, with the Aggregated Responses, from Christchurch, NZ

	Percent Who Agree or Strongly Agree with the Statement (by year)		
	2008	**2010**	**2012**
I have confidence in the council	23.9	53.5	32.0
I understand how decisions are made	34.1	37.7	24.8
I want more say in council decisions	58.2	45.8	58.2

($p < .05$). Following the 2011 earthquake, we expect this number to rise again, as people sought out more opportunities for input. In fact, the percent of people agreeing with the statement did increase to 58.2% in 2012 ($N = 559$, $M = 3.58$), indicating that more people want to be involved in the process compared to 2010 responses ($p < .05$).

Table 7.3 provides the percent of respondents indicating agreement or strong agreement with each of the three items, confidence in council, understanding how council makes decisions, and wanting more say in council decisions. For the first two measures, confidence in council and understanding how council makes decisions, there was a peak after the 2010 earthquake, followed by a dip in 2012. However, for the final measure, wanting more say in council, the result was opposite. Respondents indicated that they did not desire as much say in council following the 2010 earthquake. However, by 2012 the degree of agreement with the question increased again.

Disaster Effects on Opportunities to Participate

Opportunities for civic participation shrank with the CERA and CCDU takeover. Edwards (2009) highlights why cooperation between governance agencies, particularly ones with a vertical relationship, is marked with tension. The relationship between CCC and CERA quickly transformed from cooperative to hierarchical, with CERA the primary decision maker. Koliba et al. (2010) show how a supplanting government agency like CCDU, with limited local ties, will struggle to include the input of local bodies. CCDU operated without the informal ties and network trust established by long-standing working relationships. This means that information sharing broke down; local community organizations were no longer able to affect the decision-making process. CCDU was, by nature, insular. Without informal network ties or local accountability,

Christchurch residents no longer had access to the government. The supply of participation opportunities was effectively undermined.

Assessing how opportunities for civic participation shifted in the rebuilding years must necessarily explore both the accessibility of opportunities and the meaningfulness of those opportunities. Specifically, the *accessible entry points* component relates to both the information flow from the government (i.e., how accessible is the information) and the efforts being made by the government to collect feedback from the public (i.e., how easily citizens can identify opportunities to participate). On the other hand, the *meaningfulness of the opportunity* is concerned with whether citizen input is being used to make decisions of government, as well as citizen perception of this influence. Table 7.4 identifies the key questions asked in this study for each of these items.

In regard to the first question, "How well do citizens understand the flow of information from the government regarding the earthquake recovery decisions?" we utilize a question from the CERA Wellbeing Survey, which asks respondents to identify their "satisfaction with communication about earthquake recovery decisions" (CERA Wellbeing Survey 2014). The data for this measure are available as of 2012, and therefore only provide a post-2011 earthquake measure. In 2012, 40% of respondents indicated they were satisfied or very satisfied with the information they were receiving about recovery decisions from CERA. The responses declined significantly ($p < .05$) in April 2013 (37% were satisfied or very satisfied) and again in September 2013 (34% satisfied). The trend has continued in April 2014 (33% satisfied), although this was not significantly different from the September 2013 responses. Respondents were even less satisfied

Table 7.4 Variables Used to Assess ***Opportunities to Participate*** in Christchurch, NZ

Variable	Key Questions
Accessible entry points for citizen engagement	1. How well do citizens understand the flow of information from the government regarding the earthquake recovery decisions? 2. Are there easily identifiable mechanisms in place for citizen input into the development plan?
Meaningful opportunities for citizen engagement	1. Is citizen input being used to develop the plan for the rebuilding process? 2. To what extent do citizens believe that they influence council decisions?

with Christchurch City Council's information dissemination about earthquake recovery decisions, with only 28% satisfaction in September 2012, 31% in April 2013, 28% by September 2013, and finally 28% in April 2014 (no significant changes).

For the second question pertaining to accessible entry points for citizen engagement, the structural arrangement of the institutions is important to consider. First, the CER Act placed power at the national rather than local level, weakening citizens' ability to access government through the local mechanisms that had historically been developed in the community. Second, while CERA did host a community forum from which community voices were taken under advisement, the forum comprised only 38 members, all of whom were selected through appointment by the minister (Smart, 2014, p. 244).

The second element of opportunity is the presence of meaningful opportunities for civic engagement. The first question considered is, "Is citizen input being used to develop the plan for the rebuilding process?" Initially, the answer was yes. After the first earthquake in 2010, residents were able to rebuild as a community, and the recovery effort was led by the local government. The citizens were able to identify ways to interact with the Christchurch City Council, and citizens indicated that they felt more influential in council affairs during this period. Share an Idea had a strong, positive impact on civic engagement for a limited time.

Once the central city plan was abandoned, however, civic engagement fell dramatically as city residents' input was discarded. Carlton (2013) notes that "local dissatisfaction with the process remains high, primarily due to an overarching sense that while opinions are being solicited, they are not being afforded due consideration" (p. 10). While there was a clear opportunity for participation supplied by the government, the exercise lacked the follow-through to make it meaningful. This experience can have lasting impacts in the future calculus of citizens weighing their resources (e.g., time, money) with their desire to be involved. If people feel as though they wasted their time by attempting to participate, there may be a lack of willingness to engage in the future.

In order to examine the citizens' perception of their influence, we utilize the survey question "How much influence does the public have on council decisions?" This question comes from the New Zealand Quality of Life Survey, administered in 2008, 2010, and 2012. In 2008, 17.5% of respondents indicated that they believed the public had no influence on council decisions ($N = 491$, $M = 2.31$). As expected, this fell to 9.3% in the 2010 survey, which occurred after the first earthquake ($N = 496$, $M = 2.59$;

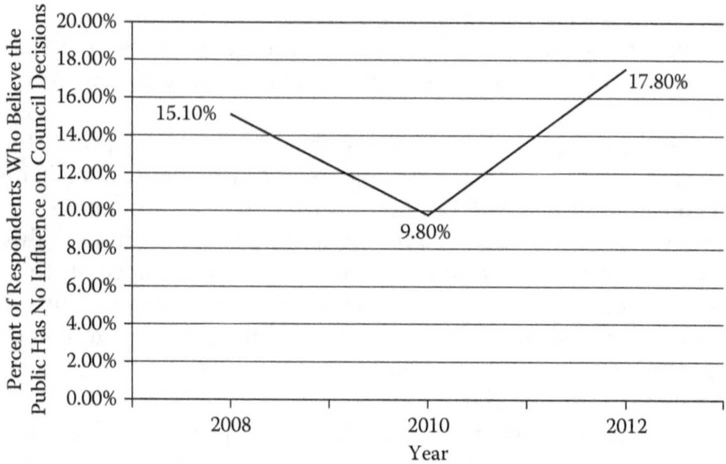

Figure 7.2 Opportunity to participate: Public influence on council.

$p < .001$). Finally, in 2012, after the shift to CERA and closure of Share an Idea, 18.9% of citizens felt the public had no influence on council decisions ($N = 529$, $M = 2.23$; $p < .001$) (see Figure 7.2). The initial rise in perceived influence in 2010, followed by a sharp decline in 2012, indicates that the shift in politics in the region had a negative impact on citizen perception of the meaningfulness of their ability to influence government decisions.

Examining Changes in Civic Engagement

In Christchurch, we expected to see a surge, followed by a fall, in civic engagement. Anecdotally, Peryman and Vallance (2012) observed the surge in civic engagement, highlighting an immense growth in community-created, community-oriented organizations in 2011 and 2012. These organizations were formed outside of, and without any support from, local government and CERA. Theoretically, we see civic engagement as an intersection of political participation, association membership, and volunteerism. As a measure of political participation, we used voter turnout (number of voters who participate in a particular election divided by the total number of enrolled voters in the district). We look at the regional turnout in both general elections and local elections. Association membership is measured by the Quality of Life Survey, which asks, "Which social networks do you belong to?" And finally, volunteerism is measured by responses to the New

Table 7.5 Voter Turnout

	Turnout—Local Elections (Year of Election)	**Turnout—General Election (Year of Election)**
Prior to 2010 earthquake	42% (2007)	81% (2008)
After 2010 earthquake, prior to 2011 earthquake	52% (2010)	Data not available (no general election in time period)
After 2011 earthquake	43% (2013)	75% (2011) 74% (2014)

Zealand General Social Survey. The survey asked if the respondent had volunteered in the past 4 weeks.

Civic engagement is the actual action taken by citizens to better their community. This may involve a multitude of activities, including the measures we have chosen to utilize: voting, membership in organizations, and volunteering. First, examining the political aspect using turnout to indicate trends in political participation, we see a decline in both the general and local election turnout rates from the time period prior to the earthquakes to that after the earthquakes (Statistics New Zealand, 2014). There was not a general election in between the events, but there was a local election (see Table 7.5). Thus, as expected, there is evidence that citizens feel less inclined to participate in the political processes. Following the entry of CERA in 2011, the political participation, as indicated by the percent of voter turnout, declined. Indeed, the last general election, held in September 2014, showed a further decline in turnout in Christchurch.

The second aspect we consider is the social aspect of civic engagement, specifically, the number of associations respondents belong to, as well as whether people volunteered in the 4 weeks prior to the survey. Figure 7.3 shows the two measures for 2008, 2010, and 2012 as measured by the average number of association, as well as the percent who indicated they volunteered in the 4 weeks prior to survey administration (New Zealand General Social Survey 2008).

As illustrated by Figure 7.3, both of our social civic engagement measures peaked in 2010 and declined in 2012. We argued that the decline in civic engagement is motivated by both the opportunity and motivation for participation. As has been shown, the interest in participation also peaked in 2010. This bolsters our suggested causal relationship; changes

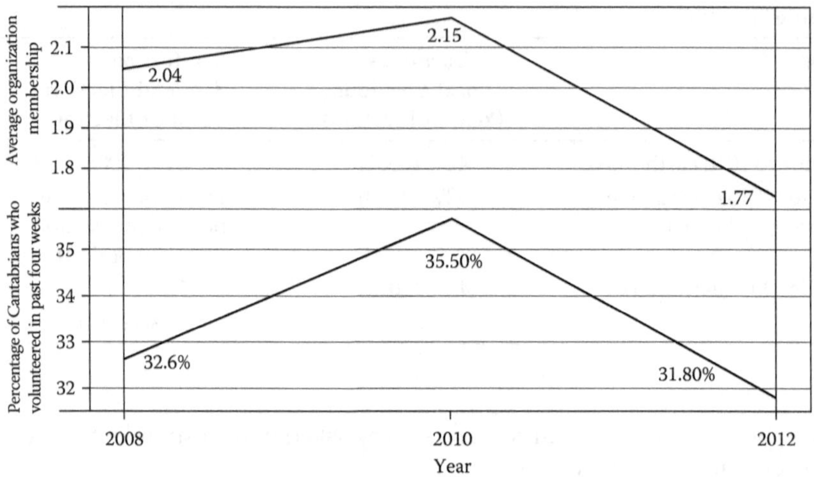

Figure 7.3 Civic engagement: Membership in organizations and volunteerism.

in opportunities to participate and motivation to participate are precursors to changes in civic engagement.

CONCLUSION

Disasters present a rare opportunity for change in the civic engagement levels of a community. In response to this type of exogenous shock, we observed that communities rally together to solve problems. We also observed communities where apathy seems to dominate, and civic engagement stagnates. Why does one community choose to engage, whereas another retracts? In this chapter, we proposed that observable civic engagement is a complex interplay between individuals' desire to participate and the institutional provision of public opportunities for frequent and meaningful interaction with local decision makers. These variables, considered together, explain the different outcomes that we saw and create lasting impacts that were underestimated and require more scholarly attention.

The framework presented here adds shades of complexity to the civic engagement field. We are aware, however, that this is a small step. Much more research is required to fully understand the interplay of variables that explain civic engagement behavior in a postdisaster context. Inclusion of socioeconomic status variables, for example, is an avenue of research that requires pursuit. Similarly, measuring the effect of preexisting resiliency on postdisaster opportunity and motivation for civic participation is worthy of attention. Our goal in this chapter was to add nuance to a discussion that has previously been underestimated; future inquiry possibilities are substantial.

REFERENCES

Adetunji, J. (2010, September 3). Earthquake strikes Christchurch in New Zealand. *The Guardian* (Christchurch).

Adler, R. P., and Goggin, J. (2005, July). What do we mean by "civic engagement"? *Journal of Transformative Education*, 3(3), 236–253.

Aldrich, D. P. (2010). *Fixing Recovery: Social Capital in Post-Crisis Resilience*. Paper 3. Purdue Political Science Faculty Publications. http://docs.lib.purdue.edu/pspubs/3.

Aldrich, D. P. (2011). Ties that bond, Ties that build: Social capital and governments in post disaster recovery. *Studies in Emergent Order*, 4, 58–68.

Aldrich, D. P. (2012). *Building Resilience: Social Capital in Post-Disaster Recovery*. Chicago: University of Chicago Press.

Bennett, A. (2011, February 24). Christchurch earthquake: Wide Powers in govt hands for national emergency. Retrieved November 1, 2014, from http://www.nzherald.co.nz/nz/news/article.cfm?c_id = 1&objectid = 10708334.

Berke, P. R., and Campanella, T. J. (2006). Planning for postdisaster resiliency. *Annals of the American Academy of Political and Social Science*, 604(1), 192–207.

Berry, J. M., Portney, K. E., and Thomson, K. (1993). *The Rebirth of Urban Democracy*. Washington, DC: Brookings Institute Press.

Birkmann, J., Buckle, P., Jaeger, J., Pelling, M., Setiadi, N., Garschagen, M., et al. (2008). Extreme events and disasters: A window of opportunity for change? Analysis of organizational, institutional and political changes, formal and informal responses after mega-disasters. *Natural Hazards*, 55(3), 637–655.

Brookie, R. (2012). *Governing the Recovery from the Canterbury Earthquakes 2010–11: The Debate over Institutional Design*. No. 12/01. Wellington, New Zealand: Institute for Governance and Policy Studies.

Canterbury Earthquake Recovery Act 2011. (2011, April 18). Retrieved August 20, 2014, from http://www.legislation.govt.nz/act/public/2011/0012/latest/whole.html#DLM3653522.

Canterbury Earthquake Recovery Authority. (2014). *CERA Wellbeing Survey April 2014 Report*. Wellington, New Zealand: AC Nielsen.

Carlton, S. (2013). Share an Idea, spare a thought: Community consultation in Christchurch's time-bound post-earthquake rebuild. *Journal of Human Rights in the Commonwealth*, 2(2), 4–13.

Chang, S. E., Taylor, J. E., Elwood, K. J., Seville, E., Brunsdon, D., and Gartner, M. (2014). Urban disaster recovery in Christchurch: The central business district cordon and other critical decisions. *Earthquake Spectra*, 30(1), 513–532.

Chamlee-Wright, E., and Storr, V. H. (2011). Social capital as collective narratives and post-disaster community recovery. *Sociological Review*, 59(2), 266–282.

Cohen, D. (2014, January 19). Christchurch Cathedral a flashpoint for quake-hit New Zealand town. *The Christian Science Monitor.* Retrieved from http://www.csmonitor.com/World/Asia-Pacific/2014/0119/Christ-Church-Cathedrala-flashpoint-for-quake-hit-New-Zealand-town.

Cox, R. S., and Perry, K.-M. E. (2011). Like a fish out of water: Reconsidering disaster recovery and the role of place and social capital in community disaster resilience. *American Journal of Community Psychology*, 48, 395–411.

Edwards, F. L. (2009). Effective disaster response in cross border events. *Journal of Contingencies and Crisis Management*, 17(4), 255–265.

Ekman, J., and Amna, E. (2012). Political participation and civic engagement: Towards a new typology. *Human Affairs*, 22, 283–300.

Fiorina, M. P. (1999). Extreme voices: A dark side of civic engagement. In T. Skocpol and M. P. Fiorina (eds.), *Civic Engagement in American Democracy* (pp. 395–426). Washington, DC: Brookings Institute Press.

Foster-Bey, J. (2008). *Do Race, Ethnicity, Citizenship and Socio-Economic Status Determine Civic Engagement?* CIRCLE Working Paper 62. Retrieved November 1, 2014, from http://files.eric.ed.gov/fulltext/ED505266.pdf.

Fung, A., and Wright, E. O. (2001). Deepening democracy: Innovations in empowered participatory governance. *Politics and Society*, 29(1), 5–41.

Gates, C., Sachdeva, S. A. M., Steeman, M., and Mcdonald, L. I. Z. (2012, April 19). Christchurch rebuild to be led by govt. *The Press* (Christchurch, NZ).

Harrald, J. R. (2006). Agility and discipline: Critical success factors for disaster response. *Annals of the American Academy of Political and Social Science*, 604, 256–272.

Kirlin, J. J., and Kirlin, M. K. (2002). Strengthening effective government–citizen connections through greater civic engagement. *Public Administration Review*, 62 (s1), 80–85.

Klandermans, B. (2007). The demand and supply of participation: Social psychological correlates of participation in social movements. In D. A. Snow, S. A. Soule, and H. Kriesi (eds.), *The Blackwell Companion to Social Movements*, Oxford, UK: Blackwell Publishing Ltd.

Koliba, C., Meek, J. W., and Zia, A. (2010). *Governance Networks in Public Administration and Public Policy*. Boca Raton, FL: CRC Press.

Leighly, J. E. (1995). Attitudes, opportunities and incentives: A field essay on political participation. *Political Research Quarterly*, 48(1), 181–209.

Levi, M. (1996, March). Social and unsocial capital: A review essay of Robert Putnam's *Making Democracy Work. Politics and Society*, 24(1), 45–55.

Mettler, S. (2002). Bringing the state back into civic engagement: Policy feedback effects of the G.I. Bill for World War II Veterans. *American Political Science Review*, 96(2), 351–365.

Meyer, D. S. (2004). Protest and political opportunities. *Annual Review of Sociology*, 30(1), 125–145.
Ministry of Business, Innovation, and Employment. (2013). *Statement of Intent 2013–2016*. Wellington, New Zealand: Crown Copyright.
Norris, F. H., Stevens, S. P., Pfefferbaum, B., Wyche, K. F., and Pfefferbaum, R. L. (2008). Community resilience as a metaphor, theory, set of capacities, and strategy for disaster readiness. *American Journal of Psychology*, 41, 127–150.
NV Interactive. (n.d.). Client: Christchurch City Council. Retrieved August 14, 2014, from Consultation Reinvented: http://www.nvinteractive.co.nz/our-work/share-an-idea.
Parker, M., and Steenkamp, D. (2012). The economic impact of the Canterbury earthquakes major earthquakes. *Reserve Bank of New Zealand: Bulletin*, 75(3), 13–25.
Peryman, P. B., and Vallance, S. (2012). *An Inventory of Community-Led Recovery Initiatives in Canterbury*.
Putnam, R. D. (1993). *Making Democracy Work: Civic Traditions in Modern Italy*. Princeton, NJ: Princeton University Press.
Quinn, B., and Trann, M. (2011, February 22). New Zealand earthquake strikes Christchurch, killing at least 65 people. *The Guardian* (London). Retrieved from http://www.theguardian.com/world/2011/feb/22/new-zealand-earthquakechristchurch.
Schofer, E., and Fourcde-Gourinchas, M. (2001, December). The structural contexts of civic engagement: Voluntary association membership in comparative perspective. *American Sociological Review*, 66(6), 806–828.
Shaw, R. (2004). Social capital: A missing link to disaster recovery. *International Journal of Mass Emergencies and Disasters*, 22(1), 5–34.
Simo, G., and Bies, A. L. (2007, December). The role of nonprofits in disaster response: An expanded model of cross-sector collaboration. *Public Administration Review*, special issue, 125–142.
Skocpol, T. (1997). The Tocqueville problem: Civic engagement in American democracy. *Social Science History*, 21(4), 455–479.
Skocpol, T., Ganz, M., and Munson, Z. (2000, September). A nation of organizers: The institutional origins of civic volunteerism in the United States. *American Political Science Review*, 94(3), 527–546.
Smart, J. (2014). The role of post-disaster institutions in recovery and resilience: A comparative study of three recent disasters. In J. Boston, J. Wanna, V. Lipski, and J. Pritchard (eds.), *Future-Proofing the State: Managing Risks, Responding to Crises and Building Resilience* (pp. 229–249). Canberra: Australian National University Press.
Statistics New Zealand. (2012). New Zealand General Social Survey: 2012. Retrieved from http://www.stats.govt.nz/browse_for_stats/people_and_communities/Households/nzgss_HOTP2012.aspx.
Statistics New Zealand. (2014). Voter turnout. Retrieved from http://www.stats.govt.nz/browse_for_stats/snapshots-of-nz/nz-social-indicators/Home/Trust%20and%20participation%20in%20government/voter-turnout.aspx.

Stolle, D., and Hooghe, M. (2005). Inaccurate, exceptional, one-sided, or irrelevant? The debate about the alleged decline of social capital and civic engagement in Western societies. *British Journal of Political Science*, 35(1), 149–167.

Taylor, J. (2011, February 23). New Zealand earthquake caused by new fault line. *The Independent* (London).

Theiss-Morse, E., and Hibbing, J. R. (2005). Citizenship and civic engagement. *Annual Review of Political Science*, 8, 227–249.

Turner, A. (2012, February 1). Quake bus victim: "They wanted me to stop screaming." *Christchurch Star* (Christchurch).

Ursano, R. J., Fullerton, C. S., Weisaeth, L., and Raphael, B. (2007). Individual and community responses to disasters. In R. J. Ursano, C. S. Fullerton, L. Weisaeth, and B. Raphael (eds.), *Textbook of Disaster Psychiatry* (pp. 1–28). Cambridge: Cambridge University Press.

8

Managing for Resilience across Multiple Scales of Action in Joplin, Missouri

Clayton Wukich, David E. A. Johnson, and Michael D. Siciliano

Contents

INTRODUCTION

Emergency managers face challenges in managing for resilience—facilitating the community-wide anticipation of, response to, and recovery from low-probability, high-impact events. These managers are expected to empower appropriate levels of organizational and interorganizational activities informed by a common operating picture that they, personally, help to facilitate (Rosenthal et al. 2001). However, the actions of others shape and constrain their performance at multiple scales of operation and impact response and recovery outcomes. Furthermore, emergency managers have limited formal control over these actors; therefore, managing within this context represents a critical administrative problem. This chapter examines the managerial and relevant external factors that influence community resilience, acknowledging that outcomes are affected by actions both in anticipation of and in response to natural disasters. Local governments are obviously not the only actors involved in this process, and coordination across social sectors and levels of government is needed to promote community resilience (Comfort et al. 2010a).

The Joplin, Missouri, tornado of 2011, the deadliest in the United States in the past 60 years, offers a case in which to examine how emergency managers identified, interpreted, and communicated shared risk across multiple scales of action and worked together to reduce that risk prior to, during, and after the tornado. Joplin provides a critical case through which to develop a set of propositions regarding the efficacy of managerial strategies to promote the kind of resilience that leads to effective and efficient recovery. We argue that recovery is affected not just by reactive processes, but also by concerted efforts across emergency management phases.

Much attention has been paid to interorganizational coordination during disasters (Drabek and McEntire 2002). Some research has examined and modeled how organizations interact across boundaries (Birkland and Waterman 2008; Comfort and Haase 2006; Kapucu 2006);

however, studies have tended to focus on specific scales, such as the national (Schneider 1990), interstate (Waugh 2007; Kapucu et al. 2009), and local (McEntire 2002) governmental levels of operations. Our goal is to link mitigation, preparedness, response, and recovery efforts on multiple scales of action (e.g., household, organizational, and interorganizational) to demonstrate the key managerial skills used to (1) protect lives and property, (2) maintain continuity of operations, and (3) facilitate quick and efficient recovery operations.

Findings indicate that the following factors facilitated resilient performance in Joplin: (1) past planning and interorganizational exercises prior to the tornado, (2) early recognition and interpretation of risk by experienced meteorologists and emergency managers, (3) implementation of atypical communication tactics with the public, and (4) effective coordination of multiple networks, which included innovative strategies across emergency support functions (ESFs). Not all actions taken were successful, however. Both successes and failures provide the basis for a set of critical management skills proposed in the findings section and outlined in the discussion section.

THEORY

Resilience

Many researchers define resilience as the ability to recover from extreme events. In his 1988 book *Searching for Safety*, Aaron Wildavsky (1988) introduced the term *resilience* to students of public administration, emphasizing the ability of actors to recover from high-impact events. Subsequent definitions by social scientists generally maintain this deference to the safety sciences and the notion of bouncing back (see McEntire et al. 2002; Vale and Campanella 2005). For example, according to Aldrich (2012, p. 7), "resilience is a neighborhood's capacity to weather crises such as disasters and engage in effective and efficient recovery through coordinated efforts in cooperative activities." This type of definition denotes a reactive quality to resilience and a predominant focus on recovery activities.

Others have suggested resilience is more than reaction to an extreme event. Comfort et al. (2010a) illustrate a comprehensive conceptualization of resilience that is both proactive and reactive, connecting it to all phases of disaster management. "Resilience," they suggest, "is the capacity of a social system (e.g., an organization, city, or society) to proactively adapt to and recover from disturbances that are perceived within the system to fall outside the range of normal and expected disturbances" (Comfort et al. 2010a, p. 9). Resilience, therefore, represents the outcome of decisions that have intentionally or unintentionally addressed shared risk and includes prevention, protection, mitigation, preparedness, response, and recovery operations. Certain qualities across scales of action were found to promote this type of resilience.

The Multidisciplinary Center for Earthquake Engineering Research, located at the University of Buffalo, has articulated a framework of resilience that considers various scales of action. According to this *R4* approach, resilient systems are

1. Robust: They withstand internal and external shocks.
2. Redundant: They possess substitutable parts that facilitate operations despite component failure.
3. Resourceful: They identify problems and develop potential solutions.
4. Rapid: They act quickly to prevent or minimize any disruptions to the continuity of critical operations or services (Tierney and Bruneau 2007, p. 15).

Factors such as the magnitude of an incident, coincidence in terms of which infrastructures are affected and which are not, and the availability of key resources influence each R4 component, and therefore incident outcomes. Other contextual factors impact community resilience. Cutter et al. (2008), in their disaster resilience model, identify multiple interdependent dimensions that make up a community's capacity for resilience, including the nature of ecological systems (e.g., their ability to absorb shock); various characteristics of social, economic, and institutional systems; and the built infrastructure. Other exogenous variables, such as whether a hazard strikes vulnerable areas, also impact recovery outcomes. In the face of these conditions, performance is enhanced by resourceful and quick-acting managers (Tierney and Bruneau 2007). Several other skills are needed, however, to ensure community resilience.

Managerial Skills

Several researchers have examined different types of management skills that facilitate resilience. Quarantelli (1997), for example, enumerated a list of traits that constitute "good emergency management," identifying both traditional management skills (e.g., the generation and delegation of tasks) and collaborative skills (e.g., interacting with the media and developing a well-functioning emergency operations center).

Comfort et al. (2010a), Aldrich (2012), and Ross (2014) have stressed the ability to coordinate as a key adaptive process, albeit at different scales of action. Aldrich (2012) links community levels of social capital, particularly the ability to gain access to public resources, with improved recovery outcomes. Comfort et al. (2010b) focused on interorganizational levels of communication and coordination, finding that managers who are able to detect and interpret risk, communicate that risk, and develop strategies for coordinated action that reduce that risk perform at higher rates than those who do not demonstrate those skills. Part of the adaptation process occurs prior to extreme events in the form of the adoption of smart architecture and prudent zoning and regulatory policies (Comfort et al. 2010b; Ross 2014).

While emergency managers retain traditional administrative responsibilities related to intraorganizational planning, operations, and human resources, increasingly they are being asked to cross organizational boundaries, recruit allies, build consensus, and coordinate action, all while scanning the external environment for shared risk (Somers and Svara 2009; Wukich and Robinson 2013). These emergent networks built by managers "enhance resilience because they raise the probability that needed information and resources will become available through network ties and because they empower even network newcomers within the context of overall response" (Tierney and Trainor 2004, p. 167).

During large-scale incidents, effective emergency management requires coordination (Drabek and McEntire 2002; Leonard and Howitt 2007; Waugh and Tierney 2007), and therefore calls for different types of skills than those demonstrated by traditional models of public administration that assume a certain level of environmental stability (Agranoff 2007). The overarching managerial skill that facilitates resilience in uncertainty conditions is the ability to respond productively to change (Horne and Orr 1997), a trait bolstered by sensemaking skills (Weick 1995) and the ability to design and implement appropriate strategies for action (Comfort 2007). Building on Comfort's work, Wukich and Robinson (2013) identify specific skills related to network performance in conditions of uncertainty, including the abilities to

1. Navigate networks
2. Recruit potential allies
3. Develop and maintain relationships

4. Acquire data to identify risk
5. Link that risk to vulnerabilities
6. Develop and execute strategies for action, often in coordination with other actors who share similar goals
7. Contribute to the formation and maintenance of a systemic common operating picture that informs effective decision making

Making sense out of uncertainty and responding productively to change requires that organizational managers transcend their administrative silos. Situational awareness informed by a common operating system facilitates these processes (Comfort 2007; Huder 2012). However, designing and implementing information and action-based workflows that enable the formation and maintenance of a common operating picture proved to be problematic. Information asymmetries, for example, inhibit evidence-based decision making and effective action (Comfort 2007).

Within this context, we ask what decisions and subsequent actions officials took to reduce risk in the immediate hours before and during the aftermath of the Joplin tornado. Our analysis considers the multiple scales of action that shaped and constrained these decisions in part by applying Harrison and Shirom's (1999) sharp-image diagnostics to examine (1) the initial set of challenges posed by the tornado and how people reacted to it (or what the authors label *system fit*); (2) the interdependencies that emerged between people and organizations, specifically with respect to the distribution of information and other resources; and (3) the gaps or inadequacies within the response system.

DATA AND METHODS

A disaster as devastating as the Joplin tornado impacts multiple scales of action and requires a coordinated response. In order to model multiple scales of action, we employed a theory-driven, embedded, single case study design. Similar to Ostrom's (2005) action arenas, the Joplin case exhibited actors who operated at multiple scales with a variety of goals, strategies, and capabilities. An examination of multiple units of analysis facilitated the assessment of this multisector, multilevel network. While a weakness of embedded single case studies is that they have the potential for an uneven focus on specific levels at the expense of the complete picture (Yin 2009), we employed Harrison and Shirom's (1999) sharp-image diagnostics and analyses to allow for analysis at the individual, dyadic, group, and system (i.e., whole network) levels.

The data for our study came from two primary sources, which documented the actions and interactions of organizations responding to the tornado: newspaper reports and Missouri state situation reports. Newspaper reports were from the *Joplin Globe* and the Associated Press (AP). The newspaper reports were identified through a LEXIS-NEXIS search of the term *Joplin tornado* (for the AP) and a detailed review of all articles published in the *Joplin Globe*. The situation reports were produced by the Missouri State Emergency Management Agency (SEMA) and offered a state-level perspective on the implementation of emergency support functions. These documents detail disaster response operations covering a 4-week span from May 22, 2011, to June 25, 2011, and include 257 newspaper articles and 38 situation reports. A 4-week period is appropriate because it captures both response and immediate recovery operations (Comfort and Haase 2006). The city of Joplin also provided us with its list of mutual aid agencies. This list helped in identifying an array of participating local agencies across 12 states. One interview with the Joplin emergency manager provided validation and direction.

The Joplin disaster operates as a critical case through which logical generalizations can be developed and subsequently applied to other cases (Miles and Huberman 1994). Throughout the study, we wove together relevant findings and theory from existing literature with data and analysis of the Joplin case to develop a set of propositions regarding the efficacy of managerial strategies during extreme events. Our analysis employed mixed methods, including content analysis of newspaper and situation reports, network analysis of the organizational response system, and hyperlink analysis of relevant websites.

Content analysis provides a systematic approach to engage texts and narratives, which allows the researcher to generate reproducible and valid inferences from the sources of data to the context under study (Krippendorff 1980). Recent work by Ikeagwuaní and John (2013) used a content analysis approach to assess fire hazards in the maritime oil sector. We developed a coding rubric made up of factors considered to be determinants of resilience by Tierney and Bruneau (2007), Comfort et al. (2010), Aldrich (2012), and Ross (2014). Relevant codes were assigned to segments of text and then analyzed collectively using the qualitative analysis software MAXQDA. Table 8.1 outlines the primary categories of our coding protocol.

The newspaper and situation reports were also used to document the interactions taking place among responding organizations. Previous work on disaster events demonstrated the usefulness of archival data contained in newspapers and after-action reports to capture the interactions

Table 8.1 Factors That Influence Resilience Identified by Newspaper Reports by Frequency

Factors	Definition	Frequency
Coordination	The degree to which relevant actors engaged in appropriate levels of coordination in order to reduce risk, protect lives and property, and maintain continuity of operations	51
Magnitude of incident	The degree to which critical infrastructure was impacted and normal patterns of action were disrupted	28
Resource availability	The extent to which resources useful to response operations were available	25
Communication of risk	The extent to which relevant actors were able to communicate risk to those affected in a timely and relevant manner	15
Interpretation of risk	The degree to which relevant actors linked risk with vulnerability	14
Detection of risk	The extent to which relevant actors detected the likelihood of the hazard	10
External regulations	The extent to which zoning and other regulatory policies addressed the risk at hand prior to impact	5
Sheer coincidence	The degree to which luck played a role, that is, identifying situations in which populations were either directly impacted by the hazard or not	4
Smart architecture	The extent to which best practices within the fields of architecture and engineering were used to mitigate against potential hazards	4

of responding organizations during extreme events (Comfort and Haase 2006; Kapucu 2006). These texts documented the actions of participants (without interference from the researchers), and thus potentially reduced bias that may have been associated with an interview or survey process. We identified and documented the interaction taking place between two organizations when information was exchanged, when resources were exchanged, or when they engaged in some other type of coordinated or cooperative activity related to disaster response. For example, in the following text taken from a May 24, 2011, newspaper article, we would code an interaction between Cable One, the local cable television provider, and the Empire District Electric Company:

> Cable One technicians are working alongside Empire District Electric Company in their repairs. But because the damage is so widespread, repair is difficult.

We also identified organizations operating independently. In the following text from a May 24, 2011, situation report, the First Baptist Church of Seneca was engaged in the response system, but on this particular day and for this particular activity, there was no indication of interaction with other actors in the system:

> The First Baptist Church of Seneca is using facilities as shelter.

This coding approach continued for all newspaper and situation reports, resulting in a comprehensive list of all named response organizations and the interactions taking place among them in the 4 weeks following the Joplin tornado. In total, we identified 313 organizations and 939 actions/interactions operating in the response network. As Provan et al. (2007, p. 480) stated, "Only by examining the whole network can we understand such issues as how networks evolve, how they are governed, and, ultimately, how collective outcomes might be generated." By integrating network analysis with content analysis, we were able to produce a more accurate understanding of the Joplin case.

We also evaluated the information flow enabled by website hyperlinks. Websites with response and recovery information, designed by responding organizations, provide another unit of analysis to assess how critical information was disseminated to the public and other responding organizations. Using issuecrawler.net, an analytic tool that searches websites for hyperlinks, we identified the linkages that exist between the sites associated with the www.rebuildjoplin.org website.

Harrison and Shirom's (1999) sharp-image diagnostics provided an overarching framework to evaluate and integrate the multiple data sources and methods by categorizing our data and analysis into three interdependent aspects of the response system: (1) the system's fit with its external environment, (2) the structure and nature of organizational interdependencies, and (3) the location of system gaps. Each of these three areas is explored in the "Results" section. In the "System Fit" section, we evaluated system fit by examining the strain the tornado put on the community and how responders worked to alleviate that strain. In the "Mapping Interdependencies" section, we identified the structural roles played by particular actors in the network (i.e., brokerage and closure positions) who facilitated systemwide response and initial recovery. In the "System Gaps" section, we investigated the repercussions of the preexisting response capacity in Joplin as well as the implications of initial decisions taken by both households and organizations on system performance.

RESULTS

System Fit

Assessing a system's ability to cope with its external environment (e.g., the initial signs of risk, the magnitude of the tornado's destruction, and the well-being of the population) represents the first step in Harrison and Shirom's (1999) sharp-image diagnostics. On Sunday, May 22, 2011, an EF-5 tornado—the Enhanced Fujita Scale's (EF Scale) most severe category for tornadoes with wind speeds estimated in excess of 200 mph—tore a broad path through Joplin, Missouri, a city of 50,000 people. A National Weather Service meteorologist described the phenomenon as "a fist coming out of the sky" (Letner 2011). The tornado killed 162 people and sent 1,371 Joplin residents to hospitals throughout the four states in the region (Missouri, Oklahoma, Kansas, and Arkansas). Casualties exceeded those of any other tornado in the United States in the previous 58 years. Only Flint, Michigan's, tornado of 1953 was deadlier. To date, rebuilding has cost $2.8 billion, making the Joplin tornado the most expensive in U.S. history (Christie 2013).

The damaged area stretched 22 miles long and up to 1 mile wide, decimating entire city blocks. Joplin's critical infrastructure—hospitals, electrical grid, telecommunication system, and transportation routes—sustained heavy damage. Over 30% of the city's built environment—8,000 buildings—was damaged or destroyed, and 30% of the population was displaced (FEMA 2011). The Joplin Chamber of Commerce estimated 300 businesses were demolished by the storm and up to 4,000 jobs were directly disrupted (Stefanoni 2011). The nearby village of Duquesne also experienced significant damage and required relief efforts.

The devastation placed heavy burdens not just on local emergency managers, but also on actors from across social sectors and levels of jurisdiction and forced a massive mobilization effort. Two sets of actions, the early detection and communication of risk and timely coordinated action, enabled emergency managers to adjust to the rapidly changing conditions.

Early Recognition, Interpretation, and Communication of Risk

In the immediate hours before the tornado, meteorologists at the National Weather Service's Springfield station and the Joplin–Jasper County

Emergency Management Agency director effectively employed their expertise and sound judgment to detect, interpret, and communicate risk to the public at large. A meteorologist tracking the storm recalled, "It was clear that a hook echo was forming and that a large tornado was developing. We picked up the debris ball here on our radar. When you see that, it signifies that major damage is going on" (quote from a *Joplin Globe* Staff Writer 2011). The Joplin emergency manager, in contact with the Springfield station, also assessed significant risk and decided to sound the warning sirens not once but twice, an atypical act since sirens are typically activated just once. This out-of-the-ordinary communication led some residents to grasp the severity of the situation. Following the disaster, the National Weather Service determined that many people failed to take immediate life-saving actions until they detected "additional extraordinary risk signals," including the Joplin emergency manager's second siren alert, as well as media instructions to take cover, or directly observed the tornado (National Weather Service 2011, p. 8).

Considering that the National Weather Service, the local emergency manager, local media, and individual citizens all participated in some way, the recognition and interpretation of risk occurred at multiple scales of action and included the synthesis of data from more than one source. At the household level, many people required a secondary source of information to convince them of the severity of the impending event. This is consistent with previous research that indicates that an initial warning will prompt additional information seeking in order to validate that warning (Tierney et al. 2001).

Proposition 1a: The recognition and interpretation of risk is an iterative process facilitated by training, experience, and judgment. This iterative process was modeled in the observe, orient, decide, act (OODA) loop framework (Huder 2012). The action arena in a disaster demonstrates many of the same conditions.

Proposition 1b: Out-of-the-ordinary risk communications or multiple point contacts may be required by emergency managers to effectively convey extreme risk signals to the community.

Multiagency Response and Coordination

The massive devastation wrought by the event far exceeded individual capabilities of households and organizations to effectively protect life and property and maintain continuity of operations. Cooperation and coordination were therefore required. Local, state, and federal disaster declarations were quickly made and external resources were mobilized, in part due to the clear communication of needs by local city and county officials. The resulting interorganizational networks provided measures to evaluate the collective decisions and actions made in the tornado's immediate run-up and aftermath.

In the minutes following the tornado, the Joplin emergency manager requested assistance from the regional incident support team comprised of public safety personnel from across southwest Missouri. They responded and rendered aid not only in the emergency operations center, but also throughout the community (personal communication 2012). Media coverage also conveyed the devastation and spurred many organizations to self-dispatch or deploy to the scene without a formal mutual aid request. Missouri state officials and the Federal Emergency Management Agency (FEMA) quickly mobilized relief efforts, operating within the framework of the Stafford Act. Graph 8.1 demonstrates the speed at which organizations arrived on scene. According to *Joplin Globe* and AP articles and SEMA situation reports, 25 organizations responded within the first hours following the tornado, 55 on the second day, and another 55 organizations on day 3. New organizations continued to arrive and provide assistance up to June 14. The fact that new organizations continued to enter the system weeks after the event demonstrates the variety of needs requiring attention during the early recovery process.

Most organizations responded, however, within the first few days of operations. The efficacy and rapidity of deployment were due, in part, to formal planning and past exercises. Less than a week prior to the tornado, Missouri state officials had exercised Emergency Management Assistance Compact (EMAC) procedures within the context of a regional planning

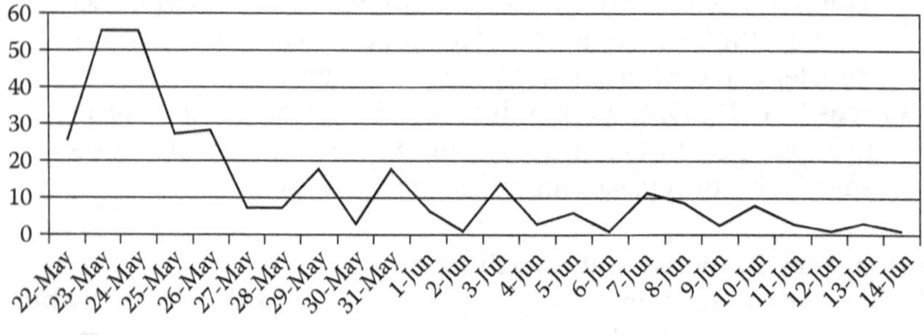

Graph 8.1 Number of organizations to enter response operations by date.

and exercise event sponsored by the Department of Homeland Security and FEMA. The National Level Exercise 2011 (NLE 11) simulated an earthquake scenario in the region and facilitated interaction between federal, state, city, and other officials in southwest Missouri. FEMA's after-action evaluation suggests that response operations in Joplin benefited as a result.

> When the tornado struck Joplin only 3 days after NLE 11 concluded, officials employed the resources, systems, and procedures that they had used in the functional exercise.... NLE 11 and other periodic exercises ensured that Joplin officials knew which regional assets were available and how to activate and employ them in the most expeditious manner. (FEMA 2011, p. 20)

According to the Joplin emergency manager, the creation of these key networks relied on trust built up among the partnering agencies, through familiarity over time. This process occurred through exercises, past incidents, and informal interaction in planning committees and professional associations (personal communication 2012).

Proposition 2a: The communication of risk and need to external partners is critical in mobilizing outside personnel and resources. This is consistent with previous research on intergovernmental relations during disasters (Waugh 2007; Comfort 2007).

Proposition 2b: Communication and interaction in accordance with the appropriate legal framework (e.g., the Stafford Act, EMAC, and local-level mutual aid agreements) prior to an incident facilitates more effective mobilization of external personnel and other resources. At the local level in this case, the familiarity among the counties in the region and associated partnering organizations proved critical in ensuring a prompt and effective response. This familiarity occurred through the planning process and experience gained by working together during previous events and informal interaction.

Proposition 2c: Regular planning and training exercises facilitate the comprehension and application of official procedures, as well as identify external resources and partners for future action.

Managing within a Diverse, Multiagency System

Organizations responded to the incident from across different social sectors and levels of jurisdiction. Table 8.2 presents the distribution of responding organizations. According to newspaper articles and SEMA situation reports, 314 organizations participated in response and recovery

Table 8.2 Frequency Distribution of Organizations Involved in Joplin Response and Recovery Operations by Social Sector and Level of Jurisdiction

	Nonprofit		Private		Public		Total	
	N	%	*N*	%	*N*	%	*N*	%
International	1	1.02	0	0	0	0	1	0.32
National	9	9.18	18	23.38	25	17.99	52	16.56
State	10	10.20	4	5.19	59	42.45	73	23.25
Regional	70	71.43	55	71.43	4	2.88	129	41.08
County	0	0	0	0	16	11.51	16	5.10
School District	0	0	0	0	5	3.60	5	1.59
Municipal	8	8.16	0	0	30	21.58	38	12.10
Total	98	31.21	77	24.52	139	44.27	314	100.00

Source: Data from Associated Press, *Joplin Globe*, and SEMA situation reports.

activities. Table 8.2 demonstrates the relatively even distribution of organizations across social sectors. Government agencies comprised 44.27%, while nonprofit (31.21%) and for-profit organizations (24.52%) contributed significantly. This relatively even distribution contrasts with Comfort and Haase's (2006) findings regarding Hurricane Katrina response networks, in which public agencies dominated the response, thus not fully engaging the collaborative capacity of the nonprofit and private sectors. This is consistent with Abramson and Culp's (2013, p. 2) finding that Joplin's "business and faith-based communities were well-established partners with each other and with local government." Whether they were large big-box stores, locally owned small businesses, or religious congregations, Joplin's business and religious organizations contributed mightily to response and recovery operations.

Within the public sector in Joplin, the local emergency management agency and fire, police, emergency medical services, and public works had developed robust mutual aid relationships with outside jurisdictions prior to the incident that were not documented by the media or state situation reports. Joplin's request for aid was answered by agencies from a wide geographic area. A large number of volunteers self-dispatched from public safety agencies, not just from the four-state area, but also from Illinois, Kentucky, and Texas. Local records indicate that more than 400 municipal and county agencies participated in the response operations in some way. Many were represented by just one or two volunteers and were not documented by SEMA situation reports or local newspaper articles (personal communication 2012). The volunteers provided a surge capacity that bolstered local personnel. The large number of volunteers and the diversity of the response system, however, presented challenges to emergency managers who were also responsible for ensuring continuity of operations and security. First, emergency managers initially lacked a comprehensive knowledge of who was available and what skill sets they exhibited. Second, in a city that established guarded perimeters to prevent looting and disaster tourism, it was difficult to sort self-dispatching volunteers from spectators.

The fact that many of these personnel and their agencies do not appear in state records points to a potential double-axis response operation, the first centered on the city of Joplin and the second on key state and national agencies. It also points to the robustness displayed by Joplin agencies that were able to not only maintain internal continuity of operations, but also actually ramp up their operational tempo to manage the influx of volunteers.

Proposition 2d: Managers coordinate among established, planned networks of actors as well as emergent, unplanned networks. This proposition is well established by past research on organizations during disasters (see Drabek and McEntire 2002; Tierney and Trainor 2004).

Proposition 2e: Networks of responders form and aggregate at different scales of action (e.g., by gravitating toward local or state functions).

Proposition 2f: Resilient, robust local organizations increase their operational tempos during disasters, as opposed to contraction. This proposition is consistent with Bakker et al.'s (2012) theory of resilient networks measured by activity.

Figure 8.1 demonstrates the broad geographic distribution of mutual aid partners. Local officials depended on the abundance of these outside resources, as did state and federal personnel who assumed operational responsibilities associated with various emergency support functions. With the internal capabilities of local agencies stretched thin, Joplin relied on the abundance of resources offered by mutual aid partners and other state and federal agencies to augment operations.

In Figure 8.1, black dots represent local jurisdictions from which mutual aid providers came; the majority resided in Missouri or within the immediate four-state area. However, the city of Joplin directly coordinated with agencies as far away as northern Illinois and southern

Figure 8.1 Geographic distribution of Joplin's local mutual aid partners.

Texas. Furthermore, according to situation reports, federal and state agencies initiated contact and facilitated coordination with teams from the distant states of California and Virginia, among others. At multiple levels of government, agencies acted as brokers—they reached out and recruited participants. At the state level, using EMAC protocols, Governor Jay Nixon's office initiated formal mutual aid requests and received assistance from several other states, including Illinois, Kansas, and Oklahoma (FEMA 2011).

Proposition 2f: In well-functioning systems, managers from across levels of jurisdiction and social sectors will recruit additional participants based on need (Wukich and Robinson 2013).

Proposition 2g: Prior understanding of mutual aid procedures—whether local, regional, or interstate—facilitates timely and efficient mobilization of external resources.

The next section explores the interdependencies and patterns of interaction created during response and recovery operations as the various organizations interacted with one another. Organizations either coordinated their efforts or acted independently to ameliorate the impact of the tornado. Organizations not only responded to the devastation, but also were forced to react to subsequent risks, such as the potential for looting, the rumor of asbestos pollution, and exposure to lightning strikes and other extreme weather.

Mapping Interdependencies

Following the assessment of system fit, Harrison and Shirom's (1999) second step in sharp-image diagnostics was to map the response system's interdependencies. This step is used to identify which actors depended on others for critical resources and the interactions that occurred in practice. Problems caused by the storm stretched across social sectors and professional disciplines. Several points of interaction enabled responders to address those problems. This section examines patterns of interaction between agencies and identifies key actors on which others depended for information and resources. We also evaluate how communication and coordination led to problem solving and adaptation. Information and communications technology (ICT) were particularly useful with respect to linking the interorganizational and household scales of action.

Multiagency Networks

In Table 8.2, we identify the number and types of organizations involved in response. Figure 8.2 demonstrates the patterns of interaction created by these organizations by sharing information and coordinating action. Isolates (organizations with no relationships) and pendants (organizations with only one link to the most active organizations) were removed to focus on core participants. National, state, and Joplin municipal agencies as well as the Joplin area school district played central roles in integrating the network. Figure 8.3 and the legend differentiate the organizations by level of jurisdiction.

Some agencies served the role of network brokers; they distributed information and other resources across social sectors and levels of jurisdiction, linking otherwise disparate groups through either directly connecting other actors via coordinated actions or facilitating important critical information flows. Other agencies maintained distinct closure patterns, working closely with similar organizations that engaged in similar work. Often agencies interacted while working on the same emergency support function. These findings are consistent with past research on brokering and closure in emergency management settings (Andrew and Carr 2013; Wukich and Robinson 2013).

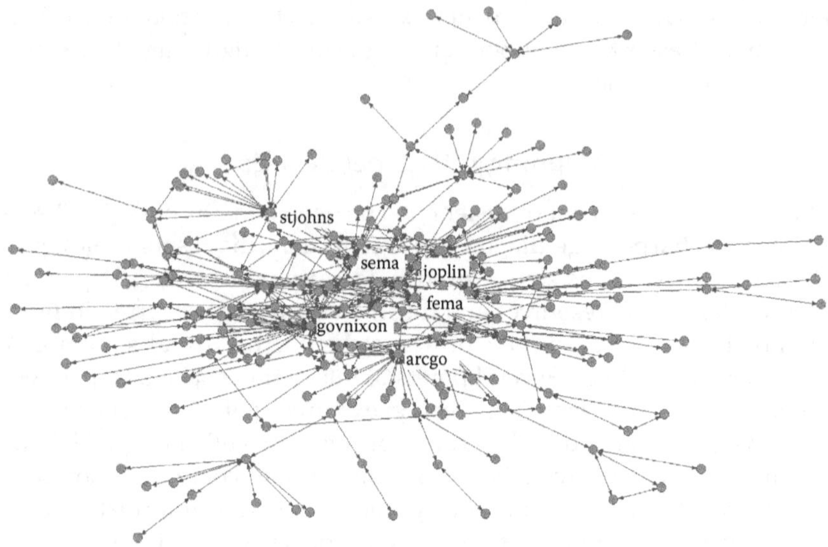

*Isolates and pendants removed.

Figure 8.2 Diagram of interacting organizations with selected actors labeled, Joplin tornado, May 22–June 15, 2011.

Figure 8.2 identifies six organizations most active in response and recovery activities: the Missouri governor's office, SEMA, FEMA, the regional chapter of the American Red Cross, St. John's Medical Center, and the city of Joplin. Each entity contributed in terms of macrolevel communication and coordination related to emergency management (the emergency support function that FEMA refers to as ESF 5) and the economic redevelopment efforts associated with long-term recovery (ESF 14). They facilitated initial response, but stayed for the duration of the recovery process to organize activities.

Table 8.3 lists organizations by degree centrality, or the total number of other organizations with which they interacted. This table statistically reaffirms the prominent roles played by the governor's office, the Red Cross, SEMA, FEMA, and the city of Joplin, organizations that represented both public and nonprofit sectors, as well as multiple levels of jurisdiction. The *Joplin Globe* is included as part of the response system because it participated in the targeted dissemination of public information.

In addition to ESF 5 and ESF 14, five other emergency support functions were particularly useful during the response and led to various patterns of

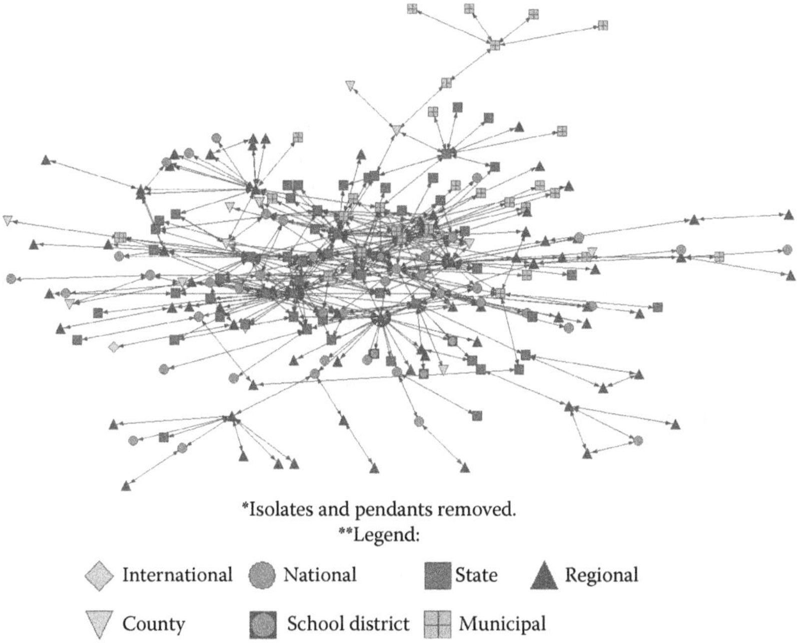

Figure 8.3 Diagram of interacting organizations by level of jurisdiction, Joplin tornado, May 22–June 15, 2011.

Table 8.3 Central Organizations Ranked by Degree Centrality

Organization	**Degree**
Gov. Jay Nixon	39
Joplin Globe	26
American Red Cross–Greater Ozarks	22
City of Joplin	21
Federal Emergency Management	20
Joplin city manager	15
Missouri Southern State University	15
American Red Cross	12
Joplin School District	12
Missouri Department of Health and Senior Services	11

interaction and mass care: emergency assistance, housing, and human services (ESF 6); public health and medical services (ESF 8); search and rescue (ESF 9); public works and engineering (ESF 3); and transportation (ESF 1).

Two nonprofit organizations played important roles as brokers, connecting otherwise disparate agencies into the central network. The first, the American Red Cross Greater Ozarks Chapter, performed central functions with regard to mass care operations (ESF 6). Red Cross officials and volunteers established shelters at Missouri Southern State University and other locations. Their predetermined role as coordinators primed to plan housing arrangements with other entities facilitated this critical position in the network. Southwestern Missouri is home to many evangelical Christian churches, for example, and the Red Cross coordinated with several during mass care operations, bringing their volunteers and resources into the larger, coordinated system.

The second nonprofit organization to play an important role as a broker was St. John's Medical Center, now Mercy Health System. Preplanning established St. John's, one of two hospitals in Joplin, as a participant in public health and medical services (ESF 8), but not as a dominant hub. However, the tornado heavily damaged St. John's facility, and the nonprofit was forced to reach out to other organizations for needed resources; for example, patients were transferred to the neighboring (and competing) Freeman Health System. St. John's worked closely with institutions from other cities and various state health officials to ensure continuity of operations for its patients and incoming victims. In cooperation with the Missouri State Disaster Medical Assistance Team, St. John's personnel quickly established a makeshift urgent care facility. This intensive workload made St. John's a bridge for many organizations to access information and other resources. For example, St. John's worked with Freeman Health System personnel to offer extensive disaster triage and urgent care, and they worked with medical insurance providers to temporarily eliminate out-of-network penalties (McClintock 2011).

Proposition 3a: Some administrative leaders pursue brokerage/bridging strategies due to their predetermined role as coordinators. Others pursue brokerage/bridging strategies in order to acquire information and other resources as the need arises.

Proposition 3b: Administrative leaders often engage in closure/bonding patterns along emergency support function lines.

Search and rescue represented another critical emergency support function in Joplin. The tornado destroyed 4,380 homes (18.8% of Joplin's housing stock) and damaged another 3,884 homes (16.7%) (FEMA 2011). In the immediate aftermath, first responders hurried to find possible survivors trapped in the debris. Firefighters, other city and county personnel, and volunteers were incorporated into these operations; hundreds of volunteers self-dispatched as well.

Information and Communications Technology

The influx of volunteers presented manpower, but they also caused managerial challenges and concerns about liability. To register and organize these volunteers, emergency managers were assisted by the St. Louis–based AmeriCorps office, which had worked with the Joplin–Jasper emergency manager during previous emergency incidents. AmeriCorps led the effort to register and credential volunteers by issuing identification cards, assigning tasks, and transporting them to and from their work locations (personal communication 2012). The information system generated by AmeriCorps and the city of Joplin helped organize what would have otherwise been a more chaotic workflow.

Proposition 4a: ICT facilitates emergent forms of human resource management.

As information technology aided emergent human resource management, information and communications technology (ICT), particularly social media, helped reunite family members and disseminate public information. Furthermore, the American Red Cross introduced a website to facilitate family reunification and many individual self-organized Facebook pages, which also reduced demands on public agencies (*Joplin Globe* Staff Writer 2011).

The devastation of Joplin's housing stock displaced thousands of families. With the destruction of landline telephones and the disruption of cell phone coverage, residents, first responders, and other organizations relied on text messaging, websites, and online peer-to-peer platforms, such as Facebook, to communicate. The Joplin school district, for example, used Facebook exclusively to communicate to students, faculty, staff, and the media in the immediate aftermath of the storm (*Joplin Globe* Staff Writer 2011).

The American Red Cross employed a central registry—www.safeandwell.com—for individuals to both list themselves as safe and search for others. There were also a number of emergent websites established to address emergent problems. For recovery activities, these sites helped link those in need with relevant support agencies and also provided reliable information for those wishing to volunteer (personal communication 2012).

In previous disasters in which responders and survivors did not have access to social media and readily built websites, activities such as reuniting families and disseminating information were often workplace based and facilitated by paid personnel, using a central phone registry (Haddow and Haddow 2014). The mobile nature of these ICT systems employed in Joplin allowed individuals to communicate with family members from hospitals, hotels, and other locations across the area. These emergent uses of ICT allowed for first and second responders to be assigned to other operations beyond traditional administrative activities.

Proposition 4b: ICT is increasingly used to disseminate critical information to residents, nonprofits, emergent volunteers, and responders during disasters.

Proposition 4c: ICT is increasingly used to facilitate self-organization during an incident.

Proposition 4d: ICT use will proliferate as more people obtain personal handheld devices with increasingly sophisticated applications and if relief organizations continue to provide Wi-Fi as a standard resource.

As focus shifted from disaster response to long-term community recovery, organizations employed ICT to disseminate information to those rebuilding. A Joplin-based nonprofit, Bright Futures, created a website—www.rebuildjoplin.org—to serve as an information hub, posting available information from a variety of sources. This website helped distribute a variety of messages, including information regarding public health warnings, temporary unemployment benefits, small business loans, and school-related scheduling; it was also used as a management tool to coordinate volunteers.

Figure 8.4 identifies the websites associated with www.rebuildjoplin.org, the vast majority derived from government agencies. Using IssueCrawler, an online analytic tool that searches websites for hyperlinks, we identified the linkages that existed between the sites. Many public agencies, particularly the governor's office and the Missouri Office of Administration, helped aggregate public information. The state also compiled a comprehensive list of links from federal and state agencies via SEMA's website. All of these sites and their linkages created a transparent information network available to organizations and residents involved in the rebuilding process.

Figure 8.4 demonstrates how government agencies such as the Small Business Administration and the Missouri Department of Agriculture

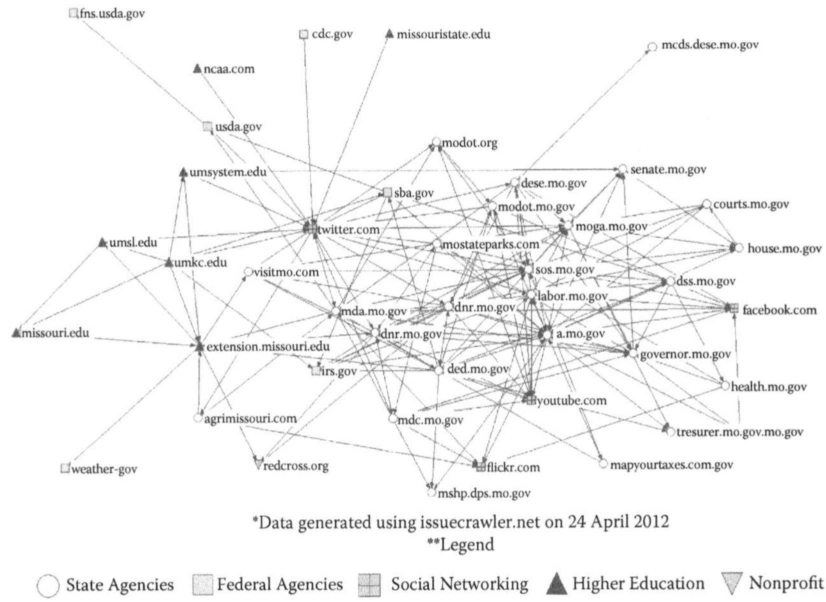

Figure 8.4 Hyperlinked websites associated with Joplin recovery efforts.

disseminated information via social media sites such as Facebook, Twitter, and YouTube to connect public goods and services to the household level of operations. This behavior is consistent with past research on the use of social media ICT during disasters and their role as information hubs to both residents and first responders (Sutton et al. 2013). This information network, amplified through social media outlets, helped spread information and empower recovery efforts.

System Gaps

The third and final step in Harrison and Shirom's (1999) sharp-image diagnostics is the identification of system gaps (e.g., problems within the system). By locating these problems, policy makers are better able to design and implement more effective strategies to recognize risk and coordinate action across multiple scales of operation.

Despite the dedication and innovation of survivors, first responders, and other organizations, some poor judgment was exercised. At the household level, decisions to ignore warnings increased casualties, which added stress to the response system. At the organizational level, communication breakdowns negatively affected specific performance requirements. Also, existing building stocks were grandfathered in when zoning ordinances were updated, exempting them from the requirements; therefore, some residents were left more vulnerable than others. Finally, at the interorganizational level, organizations working independently or within small groups apart from the core response network created asymmetry between themselves and organizations that possessed useful information resources.

While the tornado's sheer magnitude and its direct path through the city were beyond the city's control, the design of certain buildings that lacked safe rooms and storm shelters contributed to the fatality count (Paul and Stimers 2012). While this is worrisome, it is not unique. In a recent national poll, only 23% of respondents believed a natural disaster in their area was likely or very likely, and only 36% had personal/family plans in place (Zogby 2013); therefore, people were less likely to recognize risk and take appropriate action.

Household

Prior to the tornado touching down, many residents ignored the sirens even after the Joplin emergency manager took the unusual step to sound them twice. People residing in areas vulnerable to tornadoes may be desensitized to tornado warnings due to their frequency in those areas. Despite the technological ability to target siren alerts to specific geographic tracks based on weather predictions, many emergency managers continue to sound alarms for entire municipalities and counties. One meteorologist critiqued the system, "We are inadvertently training people to ignore the sirens. And that's dangerous" (McClatchy-Tribune News Service 2011).

Organizational Level

Two organizational-level problems that contributed to suboptimal outcomes in Joplin were the reliance on antiquated sirens (not unique to Joplin) and building codes that permitted grandfathering of older buildings, which did not adequately address the risk of tornadoes. The communication of risk, especially the means used to communicate, represented an initial gap between the expectations of public officials and the norms of the population. Specifically, sirens as a primary means of communication represented a problem. In light of their inadequacies, many researchers and practitioners criticized the use of sirens for public warning for tornadoes or other hazards (Hammer and Schmidlin 2002). Sirens are not always audible outside and are not intended to be heard inside, so there are large segments of the population that cannot hear the warning. This is a much larger problem that transcends Joplin.

Another problem associated with sirens is the high frequency at which they are used. Many counties and municipalities are not able to "microtarget" specific geographic areas; as a result, sirens sound for sections of jurisdictions when the threat may be more localized. Populations become desensitized to the siren, and this desensitization is seen in the increased reaction to the second siren activation, which broke the typical pattern and consequently resulted in more people paying attention.

Other governmental entities have explored and implemented equipment that allows for more precise risk communication. The advent of the NWS wireless emergency alerts, accepted by more than 100 carriers (FEMA 2013), should make a major inroad in alerting more people. In addition, the traditional allocations that were devoted to funding the sirens could be redirected to local efforts, such as reverse 9-1-1 or cell phone warnings. The application of technology is not yet comprehensive, since there are still segments of the population that do not own cell phones. Some of these segments are the more vulnerable cohorts—the elderly and the poor.

Joplin's building codes may also have contributed to suboptimal outcomes. The city of Joplin, like many others, used established building codes based on when the building was built. In the case of Joplin, the city adopted an updated version of the International Building Code in 2008. Those updated codes, however, did not apply to existing structures, only to new construction, allowing homeowners to avoid costly updates, specifically to their structure's wind load factor and strapping. The area hit by the tornado happened to be one of the city's older sections, where the structures were more fragile, and very few structures withstood the impact.

In August 2011, Joplin's building code was amended to require that new construction use hurricane straps and bolster wall attachments (Municode 2014). These should protect the structures through an upper EF-2 range (111–135 mph). Thousands of existing structures are still vulnerable, but leaders often strike compromises when implementing policies that result in high costs to residents and business owners.

Interorganizational Level

In the interorganizational network, several organizations, especially nonprofit community groups and for-profit organizations, operated independently without any reported interaction with the core group of responders. These organizations were more likely to lack access to information and other useful resources, and this lack of access points to an asymmetry between relief efforts and vulnerable populations. Unlike the key agencies or clusters of emergency support functions discussed earlier, the disconnected and periphery organizations may not have performed to their potential, depending on their access to resources.

Proposition 5a: Disconnected or peripheral organizations lack access to information and other resources.

Proposition 5b: Resilient, robust response networks incorporate as many actors as possible with respect to the flow of information and other resources. Policy makers and administrative leaders strengthen these networks both prior to and during an incident through direct interpersonal means, through indirect dissemination of information through the news media, or via ICT systems.

A specific example of information asymmetry was found in the first responders, who continued response and initial recovery activities despite severe weather warnings. When a disaster strikes, the world is not put on hold; threats and other emergent conditions may arise. In the case of Joplin, additional thunderstorms and associated lightning strikes hit the area, and a Joplin police officer was killed by one of those lightning strikes (Lehr 2011). The hot and humid weather conditions continued to impact response and recovery efforts, and these conditions increased the demand for hydration and elevated the concern for heat-related safety.

Another information dissemination problem occurred because of the destruction of several tornado sirens by the tornado. When additional tornado watches and warnings were posted several days later, mobile sirens were used to notify the public.

One strategy that greatly enhances information exchange is more robust partnerships between first response agencies and the National Weather Service. Once a tornado warning is issued, all public safety responders are notified, and in the case of the fire service, units are dispatched to predesignated observation positions. They are in radio contact with other agencies and their dispatch center, as well as emergency operation centers. Being exposed and vulnerable to weather conditions often puts response personnel at increased risk; for example, in 2008, a volunteer firefighter was storm spotting in Jasper County, just south of Joplin, when he was killed in a thunderstorm (personal communication 2012). Most public safety responders are trained in storm spotting, but these storms are treacherous. Advanced interpretation of risk conditions by National Weather Service personnel communicated on a regular basis would provide first responders with more complete information.

DISCUSSION

While historic in its damage, Joplin was not an isolated event. Similar tornadoes in more densely populated areas could have greater catastrophic effects (Simons and Sutter 2011); therefore, the key tasks for policy makers and administrators are to design and implement policies that foster community resilience—the anticipation of, response to, and recovery from low-probability, high-impact events. Joplin sits in a geographic area referred to as Tornado Alley, so-called for its susceptibility to the meteorological phenomenon of super-cell thunderstorm formations that lead to tornadoes. Because this susceptibility is a known risk, both individual households and organizations should be informed of how to better prepare and heed warning signals. Mitigation and preparedness activities represent the responsibilities of individuals and organizations alike across social sectors and levels of jurisdiction. Coupled with strong response management practices and informed recovery operations, cities and regions can absorb and rebound from these external shocks.

Informed by past research and the Joplin case, our analysis illustrates certain qualities of resilient organizational and interorganizational management. Specific management practices exist within resilient communities. Resilient managers are likely to

- Seek out and evaluate multiple sources for information
- Engage in a continual and iterative process of risk recognition and interpretation
- Communicate findings to the public and other organizations, thus contributing to a larger common operating picture
- Increasingly employ information technology, specifically information communication technology, to more effectively communicate across scales of operations
- Change planned response strategies and be willing to adapt depending on the context of the incident
- Understand the legal and administrative frameworks associated with intergovernmental cooperation
- Navigate networks of people and organizations in order to either distribute assistance or request needed materials and manpower

- Pursue brokerage/bridging or closure/bonding network strategies depending on their goals in situations
- Increase their operational tempos despite adversities
- Balance hierarchical, command and control coordinating strategies with more horizontal mechanisms depending on the context of the incident

These actions should not take place just in reaction to extreme events, but also in anticipation of them. Resilience requires forethought and action across the various phases of emergency management. Additional research should consider the relationships between these actions and the levels of resilience and subsequent performances across social sectors and levels of jurisdiction.

In the United States, federal policy recently shifted to focus more on local problems, not just on the state and federal problems. Presidential Policy Directive 8, for example, states that the systematic preparation for a variety of threats requires consultation with people and organizations across levels of government and social sectors (Obama 2011). It encourages the development of catastrophic as well as specific local risk planning, rather than the narrow focus on national planning scenarios as part of the National Preparedness Goal (Gall and Cutter 2012). Preparedness and planning, while often not considered when discussing resilience, matter because they play a role in reducing the impact of the event.

Due to the uniqueness of each situation and the diversity in term of actors and their capabilities, there is no "cookie-cutter approach" or one set of indicators (or threshold points to interpret those indicators) to ensure the proper management of risk over a distributed system of actors. Strategies to reduce risk are likewise context based; as a result, generic checklists and even carefully constructed plans fail to anticipate various contingencies in large-scale incidents (Leonard and Howitt 2007; Klein 1998). Thus, emergency managers are left to make sense of the situation and improvise. The modeling of various scales of action, however, allows emergency managers to consider various contingencies and plan for response activities in the event that an incident does occur. Understanding the potential interaction between a possible hazard and a community's level of risk and vulnerability bolsters preparedness and resilience. For emergency managers, these interactions provide essential elements of information, which are used to develop key situational awareness points (Huder 2012).

CONCLUSION

The ability of emergency managers to facilitate the communitywide anticipation of, response to, and recovery from an extreme event is shaped and constrained by many actors. The capacity to seek out pertinent and at times diverse sources of information allows emergency managers to not just recognize and interpret risk, but also adjust their strategies based on changing patterns of actions and the needs demonstrated within their communities. Modeling response systems as multiple levels of operations (e.g., households, organizations, and interorganizational networks) enables the identification of the interdependencies and gaps that influence performance and provide the basis for a set of management strategies and tactics to promote resilience.

ACKNOWLEDGMENTS

Data collection and analysis for this chapter was supported by Prof. Louise K. Comfort and the University of Pittsburgh's Center for Disaster Management (http://www.cdm.pitt.edu/). We express our sincere gratitude to her. We also thank Dr. Gunes Ertan, Stephen Coulthart, Joshua Keough, and Andrew Williamson for their data collection efforts, and Dr. Andrew Prelog for his GIS guidance.

REFERENCES

Abramson, David M., and Derrin Culp. 2013. *At the Crossroads of Long-Term Recovery: Joplin, Missouri Six Months after the May 22, 2011 Tornado*. New York: National Center for Disaster Preparedness, Earth Institute, Columbia University.

Agranoff, Robert. 2007. *Managing within Networks: Adding Value to Public Organizations*. Washington, DC: Georgetown University Press.

Aldrich, Daniel P. 2012. *Building Resilience: Social Capital in Post-Disaster Recovery*. Chicago: University of Chicago Press.

Andrew, Simon A., and Jered B. Carr. 2013. Mitigating uncertainty and risk in planning for regional preparedness: The role of bonding and bridging relationships. *Urban Studies* 50(4): 709–724.

Bakker, René M., Jörg Raab, and H. Brinton Milward. 2012. A preliminary theory of dark network resilience. *Journal of Policy Analysis and Management* 31(1): 33–62.

Birkland, Thomas, and Sarah Waterman. 2008. Is federalism the reason for policy failure in Hurricane Katrina? *Publius* 38(4): 692–714.

Christie, Les. 2013. Tornado victims may face long haul with insurers. CNN, May 22. http://money.cnn.com/2013/05/22/real_estate/oklahoma-tornado/index.html.

Comfort, Louise K. 2007. Crisis management in hindsight: Cognition, communication, coordination, and control. *Public Administration Review* 67(s1): 189–197.

Comfort, Louise K., Arjen Boin, and Chris C. Demchak, eds. 2010a. *Designing Resilience: Preparing for Extreme Events*. Pittsburgh, PA: University of Pittsburgh Press.

Comfort, Louise K., and Thomas W. Haase. 2006. Communication, coherence, and collective action: The impact of Hurricane Katrina on communications infrastructure. *Public Works Management Policy* 10(4): 328–343.

Comfort, Louise K., Namkyung Oh, Gunes Ertan, and Steve Scheinert. 2010b. Designing adaptive systems for disaster mitigation and response: The role of structure. In *Designing Resilience: Preparing for Extreme Events*, ed. L. K. Comfort, A. Boin, and C. C. Demchack. Pittsburgh, PA: University of Pittsburgh Press.

Cutter, Susan L., Lindsey Barnes, Melissa Berry, Christopher Burton, Elijah Evans, Eric Tate, and Jennifer Webb. 2008. A place-based model for understanding community resilience to natural disasters. *Global Environmental Change* 18(4): 598–606.

Drabek, Thomas E., and David A. McEntire. 2002. Emergent phenomena and multiorganizational coordination in disasters: Lessons from the research literature. *International Journal of Mass Emergencies and Disasters* 20(2): 197–224.

Gall, Melanie, and Susan L. Cutter. 2012. 2005 events and outcomes: Hurricane Katrina and Beyond. In *Emergency Management: The American Experience, 1900–2010*, ed. C. B. Rubin. Boca Raton, FL: CRC Press.

Federal Emergency Management Agency. 2011. *The Response to the 2011 Joplin, Missouri, Tornado Lessons Learned Study*. Washington, DC: U.S. Department of Homeland Security.

Federal Emergency Management Agency. 2013. Wireless emergency alerts. Retrieved June 23, 2013, from http://www.fema.gov/wireless emergency-alerts.

Haddow, George D., and Kim Haddow. 2014. *Disaster Communications in a Changing Media World*, 2nd ed. Waltham, MA: Butterworth-Heinemann.

Hammer, Barbara, and Thomas W. Schmidlin. 2002. Response to warnings during the 3 May 1999 Oklahoma City tornado: Reasons and relative injury rates. *Weather and Forecasting* 17(3): 577–581.

Harrison, Michael I., and Arie Shirom. 1999. *Organizational Diagnosis and Assessment: Bridging Theory and Practice*. Thousand Oaks, CA: Sage.

Horne, John F., and John E. Orr. 1997. Assessing behaviors that create resilient organizations. *Employment Relations Today* 24(4): 29–39.

Huder, Roger C. 2012. *Disaster Operations and Decision Making*. Hoboken, NJ: John Wiley & Sons.

Ikeagwuaní, U. M., and G. A. John. 2013. Safety in maritime oil sector: Content analysis of machinery space fire hazards. *Safety Science* 51(1): 347–353.

Joplin Globe Staff Writer. 2011. Local authorities fear death toll could eclipse 100. *Joplin Globe*, May 22. http://www.joplinglobe.com/local/x433426155/Widespread-damage-reported-after-tornado.

Kapucu, Naim. 2006. Interagency communication networks during emergencies: Boundary spanners in multiagency coordination. *American Review of Public Administration* 36(2): 207–225.

Kapucu, Naim, Maria-Elena Augustin, and Vener Garayev. 2009. Interstate partnerships in emergency management: Emergency Management Assistance Compact in response to catastrophic disasters. *Public Administration Review* 69(2): 297–313.

Klein, Gary A. 1998. *Sources of Power: How People Make Decisions*. Cambridge, MA: MIT Press.

Krippendorff, Klaus. 1980. *Content Analysis: An Introduction to Its Methodology*. London: Sage.

Lehr, Jeff. 2011. Officer dies in wake of lightning strike. *Joplin Globe*. Retrieved from http://www.joplinglobe.com/news/local_news/officer-dies-in-wake-of-lightning-strike/article_78221338-4241-52ba-bbc1-9ef7087e40af.html?mode = jqm.

Leonard, Herman B., and Arnold M. Howitt. 2007. Against desperate peril: High performance in emergency preparation and response. In *Communicable Crises: Prevention, Response, and Recovery in the Global Arena*, ed. D. E. Gibbons, 1–26. Charlotte, NC: Information Age Publishing.

Letner, Josh. 2011. Along tornado's path, victims recall trauma, wonder about future. *Joplin Globe*, May 28. Retrieved from http://www.joplinglobe.com/local/x1190396915/Along-tornado-s-path-victims-recall-trauma-wonder-about-future.

McClatchy-Tribune News Service. 2011. Storm sirens draw criticism for wide-ranging warnings. *Joplin Globe*, May 2. Retrieved from http://www.joplinglobe.com/dailybusiness/x1488642463/Storm-sirens-draw-criticism-for-wide-ranging-warnings.

McClintock, Kevin. 2011. Minor medical emergencies being addressed. *Joplin Globe*, May 26. Retrieved from http://www.joplinglobe.com/local/x1612586814/Minor-medical-emergencies-being-addressed.

McEntire, David A. 2002. Coordinating multi-organisational responses to disaster: Lessons from the March 28, 2000, Fort Worth tornado. *Disaster Prevention and Management: An International Journal* 11(5): 369–379.

McEntire, David A., Christopher Fuller, Chad W. Johnston, and Richard Weber. 2002. A comparison of disaster paradigms: The search for a holistic policy guide. *Public Administration Review* 62(3): 267–281.

Miles, Matthew B., and A. M. Huberman. 1994. *Qualitative Data Analysis: An Expanded Sourcebook*. 2nd ed. Thousand Oaks, CA: Sage.

Municode. 2014. Joplin Building Code: Sec. 26-67. Amendments to the International Residential Code. Retrieved from https://www.municode.com/library/mo/joplin/codes/code_of_ordinances.

National Weather Service. 2011. *NWS Central Region Service Assessment Joplin, Missouri, Tornado—May 22, 2011*. Kansas City, MO: National Oceanic and Atmospheric Administration.

Obama, Barack. 2011. Presidential Policy Directive/PPD-8. March 30. Retrieved from http://www.dhs.gov/presidential-policy-directive-8-national-preparedness.

Ostrom, Elinor. 2005. *Understanding Institutional Diversity*. Princeton, NJ: Princeton University Press.

Paul, Bimal Kanti, and Mitchel Stimers. 2011. *Tornado Warnings and Tornado Fatalities: The Case of May 22, 2011 Tornado in Joplin, Missouri*. Boulder, CO: National Hazards Center Quick Response Grant.

Paul, Bimal, and Mitchel Stimers. 2012. Exploring probable reasons for record fatalities: The case of 2011 Joplin, Missouri, Tornado. *Natural Hazards* 64(2): 1511–1526.

Provan, Keith G., Amy Fish, and Joerg Sydow. 2007. Interorganizational networks at the network level: A review of the empirical literature on whole networks. *Journal of Management* 33(3): 479–516.

Quarantelli, E. L. 1997. Ten criteria for evaluating the management of community disasters. *Disasters* 21(1): 39–56.

Rosenthal, Uriel, Arjen Boin, and Louise K. Comfort, eds. 2001. *Managing Crises: Threats, Dilemmas, Opportunities*. Springfield, IL: Charles C. Thomas.

Ross, Ashley D. 2014. *Local Disaster Resilience: Administrative and Political Perspectives*. New York: Routledge.

Schneider, Saundra K. 1990. FEMA, federalism, Hugo, and 'Frisco. *Publius* 20(3): 97–115.

Simmons, Kevin M., and Daniel Sutter. 2011. *Economic and Societal Impacts of Tornadoes*. Boston, MA: American Meteorological Society.

Somers, Scott, and James H. Svara. 2009. Assessing and managing environmental risk: Connecting local government management with emergency management. *Public Administration Review* 69(2): 181–193.

Stefanoni, A. B. 2011. Chamber official: 4,000 jobs affected by Sunday's tornado. *Joplin Globe*, May 26. Retrieved from http://www.joplinglobe.com/local/x816761139/Chamber-official-4-000-jobs-affected-by-Sunday-s-tornado.

Sutton, Jeannette, Emma Spiro, Carter Butts, Sean Fitzhugh, Britta Johnson, and Matt Greczek. 2013. Tweeting the spill: Online informal communications, social networks, and conversational microstructures during the Deepwater Horizon oilspill. *International Journal of Information Systems for Crisis Response and Management (IJISCRAM)* 5(1): 58–76.

Tierney, Kathleen, and Michel Bruneau. 2007. Conceptualizing and measuring resilience: A key to disaster loss reduction. *TR News*, 14–17.

Tierney, Kathleen, and Joseph Trainor. 2004. Networks and resilience in the World Trade Center disaster. In *MCEER: Research Progress and Accomplishments 2003–2004*. Buffalo, NY: Multidisciplinary Center for Earthquake Engineering Research.

Tierney, Kathleen J., Michael K. Lindell, and Ronald W. Perry. 2001. *Facing the Unexpected: Disaster Preparedness and Response in the United States*. Washington, DC: Joseph Henry Press.

Vale, Lawrence J., and Thomas J. Campanella, eds. 2005. *The Resilient City: How Modern Cities Recover from Disaster*. New York: Oxford University Press.

Waugh Jr., William L. 2007. EMAC, Katrina, and the governors of Louisiana and Mississippi. *Public Administration Review* 67(s1): 107–113.

Waugh Jr., William L., and Kathleen Tierney, eds. 2007. *Emergency Management: Principles and Practice for Local Government*. 2nd ed. Washington, DC: ICMA Press.

Weick, Karl E. 1995. *Sensemaking in Organizations*. Thousand Oaks, CA: Sage.

Wildavsky, Aaron B. 1988. *Searching for Safety*. New Brunswick, NJ: Transaction.

Wukich, Clayton, and Scott E. Robinson. 2013. Leadership strategies at the meso level of emergency management networks. *International Review of Public Administration* 18(1): 41–59.

Yin, Robert K. 2009. *Case Study Research: Design and Methods*. 4th ed. Thousand Oaks, CA: Sage.

Zogby, John. 2013. Americans neither worried nor prepared in case of a disaster. *Forbes Magazine*, May 16. Retrieved from http://www.forbes.com/sites/johnzogby/2013/05/16/americans-neither-worried-nor-prepared-in-case-of-a-disaster-sunyit-zogby-analytics-poll/.

9

Vulnerabilities Magnified

A Closer Look at Disasters and Displacement

Ann-Margaret Esnard and Alka Sapat

Contents

INTRODUCTION

As populations continue to grow and migrate to urban areas, devastation caused by disasters will increase. In developing countries, disasters tend toward a higher rate of fatalities, in part due to inadequate infrastructure, lack of building codes, and poor land use. In the developed world,

> the cascading consequences of disasters increase as supply chains and critical infrastructure become more interdependent in a global economy. Combined decadal economic and insured losses to natural disasters have increased by a factor of nearly 7 since the 1980s. (NRC, 2011, p. 3)

Rapid urbanization and population growth has increased vulnerability, strain, and overload of infrastructural and ecological systems (NRC, 2006). Continuing population growth and economic and infrastructure development resulted in a concomitant increase in the magnitudes and significances of loss and disruption associated with hazard activity, as well as increased exposure in situations where the probability and intensity of hazard activity remained constant (Paton, 2006). Megacities are particularly susceptible to disasters given their social diversity, disparities in wealth, and large-scale illegal squatter settlements (Wisner, 2003; Trice, 2006). As cities and urban areas (large and small) place intense pressure on environmentally sensitive land, including fragile coastal areas, floodplains, earthquake fault zones, and steep slopes, population displacement and longer recovery time frames are becoming more commonplace as well (NRC, 2006; Paton and Johnston, 2006). Vulnerable persons with the least ability to respond to disaster impacts are also at greatest risk of short- and long-term displacement.

The Internal Displacement Monitoring Centre's (IDMC) report *Global Estimates 2012: People Displaced by Disasters* estimated that during the period 2008–2012, 144 million people in 125 countries were forced from their homes, and 32.4 million people in 82 countries were displaced by natural disasters in 2012 alone (IDMC, 2013a). That same report documented that 98% of this displacement in 2012 was due to weather-related hazards like floods, storms, and wildfires (IDMC, 2013a). While Asia and West and Central Africa bore the brunt, 1.3 million people were displaced in high-income countries, with the United States particularly affected (IDMC, 2013a). The magnitude 7 earthquake that struck Haiti in 2010 also produced what IDMC refers to as "the highest relative level of displacement experienced by any country" (IDMC, 2013a, p. 7), and according to IFRC (2012), one-sixth of the population of the capital city Port-au-Prince (approximately 500,000 people) left the capital and sought refuge in other provinces.

The first two sections of this chapter offer a review of existing scholarship on disasters, exposure in hazard-prone areas, vulnerability, and displacement. We include examples from across the globe to highlight the scope of the displacement problem. This is followed by a section that discusses the intended and unintended consequences of postdisaster/

postcrisis development in cities that have been impacted by disasters and crises. The importance of historical context is emphasized as we provide examples of repeated and protracted displacement. We then discuss how adaptive capacity needs to be considered as part of a holistic and coordinated approach to vulnerability reduction and resilience. The final two sections offer some challenges that lie ahead and concluding statements. Parts of this book chapter draw from the authors' prepublished work as indicated in the acknowledgments.

VULNERABILITY AND PREDISPOSITION TO DISPLACEMENT

> Vulnerability is the susceptibility of human settlements to the harmful impacts of natural hazards. Impacts of concern include: (i) injuries and death to human populations; (ii) damage to personal property, housing, public facilities, equipment, and infrastructure; (iii) lost jobs, business earnings, tax revenues, as well as indirect losses caused by interruption of business and production; and (iv) public costs of planning, preparedness, mitigation, response and recovery. (Deyle et al., 1998, p. 121)

Vulnerability is a multidimensional construct captured in physical/exposure, socioeconomic/human dimensions, and society's capacity to withstand disasters (Bohle et al., 1994; Dow and Downing, 1995; Bogard, 1989; Downing, 1991; Dow, 1992; Smith, 1992; Cutter, 2003). The susceptibility of human settlements to the harmful impacts of natural disasters has implications at the individual, household, and community levels and potentially harmful outcomes, such as injuries, deaths, damage to housing and infrastructure, and destruction of businesses and livelihoods.

Physical vulnerability refers to exposure of people, property, and structures, as well as locational characteristics and structural integrity. Common indicators of physical vulnerability summarized by Sapat and Esnard (2013) include the following:

- Hazard zones (e.g., flood zones/percent urbanized area in flood zones, earthquake fault lines, soils that are highly erodible and prone to landslides, barrier islands, and mature stands of natural vegetation prone to wildfires)
- Built environment: inventory and integrity
    - Building inventory—type, location, occupancy, and age (as it relates to conformance with building codes)

    - Structural integrity (e.g., percent of manufactured housing) and structural type (wood frame, steel frame, unreinforced masonry), especially for earthquake vulnerability assessments
- Critical infrastructure (roads, bridges, water supply, sewerage, electric power systems, telecommunications)—type, location, structural integrity

Socioeconomic vulnerability is more nuanced and complex and refers to the inability of people, organizations, and societies to withstand adverse impacts to hazards. Below is a listing of traditional socioeconomic indicators, also summarized in Sapat and Esnard (2013):

- Income (e.g., low income/high poverty, housing affordability)
- Economy (e.g., occupations with focus on single-sector economic dependence, employment gain/loss);
- Age and gender (e.g., elderly, children/youth)
- Disability (e.g., physically handicapped, mentally handicapped, ill, requiring oxygen or other special appliances)
- Disadvantaged (e.g., single heads of household, government-assisted households)
- Race/ethnicity (e.g., ethnic/racial/language minorities)
- Education and literacy levels
- Housing tenancy (renters vs. owners)

One factor that has become increasingly critical in the U.S. context is the failure of programs to address the composition and diversity of families and households of unrelated individuals (Phillips and Fordham, 2009). Laska and Morrow (2007, p. 16) have further noted that "social vulnerability factors are not mutually exclusive, but tend to be clustered in patterns of vulnerability that place some communities and households at particular risk," with political, social, and economic factors used to determine what land is developed, what is built, and who lives in specific locations. This unequal exposure to risk coupled with unequal access to resources (Bolin and Stanford, 1998) is critical to our understanding of vulnerability and is an important determinant in the design of programs and initiatives for mitigating, anticipating, coping with, resisting, and recovering from the impacts of disasters.

At the global level, the list of vulnerability indicators has grown to incorporate displaced persons. The website of the International Federation of Red Cross and Red Crescent Societies (IFRC, 2014) includes

- Displaced populations who leave their habitual residence in collectives, usually due to a sudden impact disaster, as a coping mechanism and with the intent to return
- Migrants who leave or flee their habitual residence to go to new places, usually abroad to seek better and safer perspectives
- Returnees—former migrants or displaced people returning to their homes

Displacement, particularly protracted displacement, results from or is exacerbated by preexisting physical, human, societal, and institutional vulnerabilities, as well as societal and institutional responses and capabilities. The poor, homeless, elderly, children, and renters are among the categories of persons who are vulnerable to displacement after a disaster, though they will likely experience this differentially given the various permutations of socioeconomic status. There are difficulties for displaced families in finding durable solutions to end their displacement. After the 2010 Haiti earthquake, those who stayed in camps for the longest periods of time were the most vulnerable and had the fewest resources to recover from the shock of the earthquake (OCHA, 2013).

Poverty and homelessness are inextricably linked, and are both causes and outcomes of displacement. While there is a fairly large overlap between those two populations, the relationship between poverty and homelessness is complex and varies regionally (United Nations, 2005; Cross et al., 2010). Preexisting societal problems, such as homelessness, housing shortages, tight government budgets, land use disputes, and inadequate lifelines, become evident after disasters as cumulative and "late-blooming impacts" of the disaster (Mileti, 1999; NRC, 1994).

Within the United States, the complex relationship between housing and poverty results in the poorest being the hardest hit during disasters, as they are the most likely to live in hazard-prone areas and in substandard housing such as mobile homes. They are also likely to disproportionately suffer the most during the recovery phase, when housing and qualification requirements continually remain a barrier to aid and assistance to the homeless and renters. According to Bolin (2007, p. 125), after the Northridge earthquake, "federal housing assistance programs were criticized for their class biases. Programs provided far less (or no) assistance to renters, the unemployed and the homeless while they provided the most generous assistance to the middle class employed homeowners." After Hurricane Katrina in 2005, housing for displaced populations, particularly affordable housing, was in short supply as rents rose considerably (Sapat et al., 2011). In the first year after Katrina, rents rose by approximately 36%, and 3 years after the storm, rents were approximately 46% higher than before the storm, with wages not keeping pace (Brookings Institution and GNOCDC, 2008). These rising rents had their greatest impact on low-income renters, who spent a larger percentage of their income on housing (Crowley, 2009). In October 2012, Superstorm Sandy affected 24 states, with major impacts felt in the states of New Jersey and New York. More than three-quarters of a million people in the United States were forced to leave their homes (IDMC, 2013a). A major problem was the availability of postdisaster housing, given the limited housing (and affordable housing) options, specifically in New York City, one of the most difficult housing markets, with low vacancy rates and high rents (Conlin, 2012; Schwirtz, 2012).

Institutional and political vulnerabilities and fragilities must be taken into consideration as well. Indicators and indices of vulnerability to disasters tend to measure the outcomes of institutions, such as the quality of housing and infrastructure and physical risk (physical vulnerability), health and human development (social vulnerability), income inequalities (economic vulnerability), and number of those displaced (displacement vulnerability). Undergirding these outcomes are the roles played by institutions, both formal (rule of law, regulations, constitutional codes) and informal (cultural norms, traditions, governance processes). Institutional type, strength, and effectiveness are responsible in large part for different levels of vulnerability across geographical areas. Increased government effectiveness has been associated with disaster risk reduction strategies, which include early warning, preparedness, and response systems (UNISDR, 2009). Within institutions, political actors play a key role. As noted in Sapat and Esnard (2013), elected officials often adopt short-term measures to satisfy constituents and groups, particularly the more politically and economically powerful ones. Engaging in disaster response is likely to yield greater political dividends than adopting less media-friendly mitigation measures. With the 24-hour news cycle, political careers may be broken or bureaucrats may lose their jobs if disaster response is seen as being inadequate. The political and economic rewards of increasing development benefiting primarily the more affluent, particularly in crowded cities, can also exacerbate existing societal inequalities and vulnerabilities, thereby increasing potential displacement.

HAZARD-PRONE URBANIZED AREAS

> Around the globe, many primate cities and "megacities"—the urban places that serve as economic engines for entire societies—are located in area with high hazard exposure. (Tierney, 2007, p. 176)

Some of the conditions that make individuals and households vulnerable are inherent, such as age and economic status, but others are societally based, particularly with urbanization of known hazard-prone areas (Chambers and Conway, 1991; Reiss, 2011). For example, La Paz, the capital city of Bolivia, was built in a known geographically unstable area, which lies in a narrow valley and is crossed by more than 200 rivers (Scandlyn et al., 2013). The rapid growth of the city forced many people to live in floodplains and steep hills, making them particularly vulnerable to floods and mudslides during the December to March rainy season (Scandlyn et al., 2013).

Ordinario (2012) reported on an evaluation study, "Asian Development Bank's Response to Natural Disasters and Disaster Risks," which singled out Vietnam, Bangladesh, and the Philippines as the countries at highest risk of human losses and economic damage. That same study reported that about 85.2% of the Philippines' economy was prone to natural disasters, and 50.3% of the country's land area was economically at risk from multiple hazards such as floods, typhoons, and earthquakes (Ordinario, 2012). The Philippines is often described as "the melting pot of disasters." Located in the tropics, it experiences typhoons, floods, and monsoonal rains on a regular basis. As part of the "Pacific Ring of Fire," the Philippines is also prone to earthquakes and volcanic eruptions (Iuchi and Esnard, 2008). Metro Manila's rapid urbanization and population growth rates since the 1950s led to increasing numbers of the population living in poverty and in substandard housing conditions. Typhoon Haiyan, which struck in December 2013, was one in a series of disasters to affect the Philippines. According to the IDMC (2013b), the Philippines faced levels of displacement approximating 4.5 million people. Such high risk to recurrent disasters also leads to recurrent and protracted population displacement.

While these regions are particularly vulnerable to disasters and large-scale displacement due to geographical features and natural geophysical characteristics, urbanization and development compound vulnerabilities to hazards. Mileti (1999) refers to the exposure of expanding capital stock to natural hazards. This capital stock is mostly in dense urbanized areas with exposed lifeline infrastructure, such as transportation networks, electrical networks, water networks, and other critical facilities, as well as vulnerable commercial and residential structures (Davidson and Lambert, 2001;

Miles and Chang, 2006). Within developing countries, urbanization is not always synonymous with investment in infrastructure and resources, but is often associated with a lack of appropriate and effective land use planning and building standards, and deterioration of older, densely packed inner-city areas, as well as increased vulnerabilities to natural disasters. Informal squatter settlements, often populated by those already displaced from rural impoverished areas, are particularly vulnerable. These informal settlements in the most hazard-prone areas are also susceptible to regular or annual displacement due to hazards they experience. For instance, Sanderson (2000) noted that in central Delhi, India, a large squatter settlement existed within the designated floodplain of the Yamuna River for a number of years and was forced to evacuate at least once a year for several weeks at a time. The disproportionate exposure of poorer families to hazards, combined with substandard housing unable to withstand predictable, smaller-scale hazard events, directly contributes to the risk of prolonged displacement when homes are destroyed or severely damaged (IDMC, 2013a).

(RE)DEVELOPMENT INJUSTICES AND HISTORICAL CONTEXT

King (2006, p. 293) noted that planning is an "indirect reducer of vulnerability, through its role in developing services, facilities, infrastructure and access," but that most of the existing vulnerable locations are the results of historical decisions about which modern planners can do little until redevelopment. Using the 1927 Mississippi flood and Hurricane Katrina in the United States as examples, Button (2009) described development projects as contributing to the formation of those disasters. In the case of New Orleans, we state that displacement risk was produced over several decades, given the increased exposure of residents to natural hazards. A combination of political, social, and economic factors determined what land was developed, what was built, and who lived in the most vulnerable areas. According to Laska and Morrow (2007, p. 16), "past social and political decisions, combined with economic decline and engineering errors, laid the groundwork for turning this hurricane into a disaster of catastrophic proportions." This viewpoint is shared by Rivera and Miller (2010), who describe minority experiences, injustices, and struggles as shaped by the many social, economic, political, and developmental systems at play and the historical "accumulations" that built them.

Similarly, when analyzing the impacts of Hurricane Katrina in New Orleans, Spence et al. (2010) contended that the neighborhoods hit hardest were not necessarily the poorest, but they were the most segregated in the city; the highest concentration of black residences was located in a less desirable location of the city, closest to the levees. Residents of the Ninth Ward in particular were severely affected by Hurricane Katrina, resulting in displacement to nearby host communities in southeastern U.S. communities. However, their initial displacement dated back to 1964, when residents of Fazendville, Louisiana, were forcibly moved to the Lower Ninth Ward as a result of land acquisition by the National Park Service.

Disasters potentially result in opportunities to redress past historical patterns and mistakes and redevelop in better ways. For instance, noting the durable capital infrastructure in well-developed cities, Siolda (2014) raised the issue of disasters as a window of opportunity for reorganization of a city's structure, perhaps at a "reduced cost." The author used the case of the historical 1906 San Francisco fire to study the redevelopment effects, and found residential density increased over 40% in areas razed by fire relative to unburned areas. The increase in housing stock served to alleviate demand pressures on the predisaster housing stock.

However, while positive impacts might result from redevelopment, it is important to note that the postdisaster period and redevelopment projects also entail redevelopment injustices for disaster survivors and displaced persons, which in turn reproduce the conditions for future risk. In observations related to recoveries from 9/11 and Katrina, Gotham and Greenberg (2014) coined the term *crisis-driven urbanization*. They contend that low-income, disproportionately nonwhite communities, workers, and small businesses, or the primary victims of disaster, were disadvantaged in receiving aid, while wealthy, disproportionately white neighborhoods and high-end industries were privileged. They trace how uneven redevelopment transformed the postdisaster city and catalyzed gentrification and displacement in low-income areas (Greenberg, 2014).

NEW REALITIES: RESILIENCE AND ADAPTIVE CAPACITY

Resilience is defined by Norris et al. (2008) as a "process linking a network of adaptive capacities (resources with dynamic attributes) to adaptation after a disturbance or adversity" (p. 127). These scholars offer the important perspective of linking a network of adaptive capacities, including economic development, social capital, information and communication, and community competence. Capacities and disaster readiness are enhanced through interventions and policies (Norris et al., 2008; Sherrieb et al., 2010). We also need to better understand how interdependencies between people, their communities, and societal institutions and organizations influence resilience and adaptive capacity (Paton, 2006). Resilience is a measure of how well people and societies can adapt to a changed reality and capitalize on the new possibilities offered.

Some cities, like San Francisco (United States) and Wellington (New Zealand), were built on fault lines that were active in historical times, and therefore remain susceptible to future seismic activity and catastrophic damage. Halting future development is not feasible or practical, and initiatives are increasingly centered around planning and building capability and capacity to confront the consequences of hazard activity, and to coexist with the potentially hazardous elements of the environment (Paton, 2006). In essence, there is increased emphasis on adaptation and adaptive capacity, described by Paton (2006) as comprising four general components: (1) resources, (2) competencies, (3) integration/interconnectedness/interdependencies, and (4) sustained access and activity.

Paton (2006) warned that planning for adaptive capacity is no easy task given the permutations that arise from the interactions between community and hazard characteristics. He adds that we need to be mindful of related challenges in low-probability areas (that is, in locations characterized by less frequent hazard activity) as part of strategies related to risk communication. Using the context of climate change, Smit and Wandel (2006) pointed out that (1) a system's adaptive capacity and coping range (one feature of capacity) is not static, (2) external socioeconomic and political factors lead to a narrower coping range, and (3) coping ranges are flexible and respond to changes in economic, social, political, and institutional conditions over time. This differential vulnerability, including the differential access to resources, and varying levels of coping capacity, should be of concern to local emergency management officials, disaster planners, and policy makers. This is in addition to identifying the areas within their communities that have population segments with the highest levels or clusters of socially vulnerable residents.

OTHER CHALLENGES

The predisaster mitigation and postdisaster recovery planning phases both hold promise for reducing displacement vulnerability within cities and other areas. However, there are several challenges. Much of what is known about displacement causation is derived from outside the field of disaster research, and from outside the United States (Oliver-Smith, 2009). This has implications for the measurement and quantification of the displaced persons in cities, as well as those at risk for displacement. Within the United States there were several algorithm and index development research efforts aimed at measuring and quantifying displacement vulnerability (FEMA, 2006; French et al., 2008; Esnard et al., 2011), but all with limitations, including (1) lack of accounting for thresholds/tipping points/criticality; (2) difficulty of addressing complexity: interdependence, compounding, and cumulative vulnerability, particularly from multiple disasters; and (3) inability to assess cumulative impacts of policy and institutional changes by agencies and organizations. At the most basic level, the selection of indicators and variables related to displacement vulnerability is affected by the uniqueness of each disaster, temporal variation (short-term or long-term impacts), differential variability in vulnerability and coping capacity across communities and households within communities, the difficulty of determining the effect (positive or negative) of various indicators on displacement, and the lack of primary data and longitudinal research on migration patterns and mover characteristics caused by catastrophic disasters (Esnard et al., 2011; Esnard and Sapat, 2014). Within urban areas, the greatest data collection constraint with regard to displaced individuals is the "invisibility factor" (Crisp et al., 2012). Displaced people choose to blend into the crowds and "disappear" as part of their survival strategy, or they may intentionally not be counted by governments that do not want to officially acknowledge them as internally displaced persons, thereby avoiding the need to provide services.

This issue is related to the implications that need to be addressed for host communities and cities that become home to disaster survivors. Earlier in this chapter, it was noted that one-sixth of the population of Port-au-Prince sought refuge in other provinces in Haiti. In the case of Hurricane Katrina, displaced persons and families were scattered throughout the United States. The reality is that host communities themselves face considerable challenges and vary in their levels of capacity and coordination in dealing with evacuee ingress (Gerber, 2010). Receiving cities, which are often already overcrowded, have to contend with a number of issues: financial constraints related to the short-term and long-term costs of medical care, social services, shelter and temporary housing, transportation, job availability and provision, and so forth. As suggested by Esnard and Sapat (2014), this problem needs to be addressed by cities and towns, and planning at the local level requires integration and coordination both vertically (with regional, state, and national policies) and horizontally with various agencies and institutions (including private and nonprofit organizations, and community leaders) to develop strategies to address these issues.

CONCLUSION

Devastating earthquakes, catastrophic hurricanes, earth-scorching wildfires, and damaging tornadoes continue to unfold in the news, as do images of families and communities reeling from the destructive impacts. Cities, particularly megacities, across the globe are especially vulnerable to population displacement, albeit differentially based on location, levels of hazard exposure, socioeconomic factors, and community capacity. Displacement vulnerability is a complex problem, and there is a need for multifaceted and multidisciplinary approaches. We need to think about joint solutions to displacement, including the repeated and protracted displacement faced by our most vulnerable residents. Viewing displacement through a vulnerability lens (i.e., physical and socioeconomic vulnerability) provides a framework by which to draw on disaster research and myriad fields, such as emergency management, sociology, anthropology, public health, planning, and policy makings. Ensuing dialog among urban planners and policy makers about predisaster entry points to mitigate postdisaster displacement (i.e., to offset displacement vulnerability) partly relies on an understanding that displacement is both a process and an outcome. It derives from preexisting and shifting physical and socioeconomic vulnerabilities and societal problems further exacerbated after disaster. Adaptive capacity holds promise, but has to be couched in coordinated strategic efforts between various levels of government, institutions, community-based advocacy groups, and residents of both impacted communities and host communities.

ACKNOWLEDGMENTS

The material in this chapter is based on research supported by the U.S. National Science Foundation (NSF Grants CMMI-0726808, CMMI-1034667, and CMMI-1162438). Any opinions, findings, and conclusions or recommendations expressed in this material are those of the authors and do not necessarily reflect the views of the National Science Foundation. We also acknowledge publishers, journals, and other entities for granting us permission to reuse small portions of our book *Displaced by Disaster: Recovery and Resilience in a Globalizing World*, published by Routledge Press, an article published in the *International Journal of Mass Emergencies and Disaster*, and course material prepared for FEMA's Emergency Management Institute.

REFERENCES

Bogard, W.C. 1989. Bringing social theory to hazards research: Conditions and consequences of the mitigation of environmental hazards. *Sociological Perspectives*, 31, 147–168.

Bohle, H.G., Downing, T.E., and Watts, M.J. 1994. Climate change and social vulnerability: The sociology and geography of food insecurity. *Global Environmental Change*, 4, 37–48.

Bolin, B. 2007. Race, class, ethnicity, and disaster vulnerability. In H. Rodriguez, E.L. Quarantelli, and R.R. Dynes (eds.), *Handbook of Disaster Research* (pp. 113–129). New York: Springer.

Bolin, R., and Stanford, L. 1998. *The Northridge Earthquake: Vulnerability and Disaster*. New York: Routledge.

Brookings Institution and GNOCDC. 2008. *Anniversary Edition: Three Years after Katrina*. New Orleans: Greater New Orleans Community Data Center. www.gnocdc.org.

Button, G.V. 2009. Family resemblances between disasters and development-forced displacement. In A. Oliver-Smith (ed.), *Development and Dispossession: The Crisis of Forced Displacement and Resettlement* (pp. 255–274). Santa Fe, NM: SAR Press.

Chambers, R., and Conway, G. 1991. *Sustainable Rural Livelihoods: Practical Concepts for the 21st Century*. IDS Discussion Paper 296. Brighton, UK: Institute of Development Studies.

Conlin, M. 2012. Sandy refugees say life in tent feels like prison. Thomson Reuters Foundation. www.reuters.com/article/2012/11/10/us-storm-sandy-tentcity-idUSBRE8A90BV2012110.

Crisp, J., Morris, T., and Refstie, H. 2012. Displacement in urban areas: New challenges, new partnerships. *Disasters*, 36(S1), S23–S42.

Cross, C., Seager, J., Erasmus, J., Ward, C., and O'Donovan, M. 2010. Skeletons at the feast: A review of street homelessness in South Africa and other world regions. *Development Southern Africa*, 27(1), 5–20.

Crowley, S. 2009. Testimony of Sheila Crowley, MSW, PhD, president of the National Low Income Housing Coalition, presented to the Ad Hoc Subcommittee on Disaster Recovery of the Committee on Homeland Security and Government Affairs, U.S. Senate. March 18. www.nlihc.org/doc/Testimony-of-Sheila-Crowley-March-18-2009.pdf.

Cutter, S.L. 2003. The vulnerability of science and the science of vulnerability. *Annals of the Association of American Geographers*, 93(1), 1–12.

Davidson, R.A., and Lambert, K.B. 2001. Comparing the hurricane disaster risk of U.S. coastal counties. *Natural Hazards Review*, 2(3), 132–142.

Deyle, R.E., French, S.P., Olshansky, R.B., and Patterson, R.G. 1998. Hazard assessment: The factual basis for planning and mitigation. In R. Burby (ed.), *Cooperating with Nature*. Washington, DC: Joseph Henry Press.

Dow, K. 1992. Exploring differences in our common future(s): The meaning of vulnerability to global environmental change. *Geoforum*, 23, 417–436.

Dow, K., and Downing, T.E. 1995. Vulnerability research: Where things stand. *Human Dimensions Quarterly*, 1, 3–5.

Downing, T.E. 1991. Vulnerability to hunger and coping with climate change in Africa. *Global Environmental Change*, 1, 365–380.

Esnard, A.M., and Sapat, A. 2014. *Displaced by Disasters: Recovery and Resilience in a Globalizing World*. Boca Raton, FL: Routledge Press, Taylor & Francis.

Esnard, A.M., Sapat, A., and Mitsova, D. 2011. An index of relative displacement risk to hurricanes. *Natural Hazards*, 59(2), 833–859.

FEMA (Federal Emergency Management Agency). 2006. HAZUS: FEMA's software program for estimating losses from disasters. www.fema.gov/plan/prevent/hazus/index.shtm.

French, S.P., Feser, E., and Peacock, W.G. 2008. *Quantitative Models of the Social and Economic Consequences of Earthquakes and Other Natural Hazards* (pp. 1–50). Final report. Mid-America Earthquake Center Project SE-2. Urbana, IL: University of Illinois at Urbana-Champaign.

Gerber, B.J. 2010. Management of evacuee ingress during disasters: Identifying the determinants of local government capacity and preparedness. *Risk, Hazards and Crisis in Public Policy*, 1(3), article 7.

Gotham, K.F., and Greenberg, M. 2014. *Crisis Cities: Disaster and Redevelopment in New York and New Orleans*. Oxford: Oxford University Press.

Greenberg, M. 2014. The disaster inside the disaster: Hurricane Sandy and post-crisis redevelopment. *New Labor Forum*, 23(1), 44–52.

IDMC (Internal Displacement Monitoring Centre). 2013a. *Global Estimates 2012: People Displaced by Disasters*. Geneva: IDMC.

IDMC (Internal Displacement Monitoring Centre). 2013b. *Philippines: Comprehensive Response to Wave of Displacement Crisis Needed*. Geneva: IDMC.

IFRC (International Federation of Red Cross and Red Crescent Societies). 2012. *World Disasters Report 2012, Focus on Forced Migration and Displacement*. Geneva: IFRC.

IFRC (International Federation of Red Cross and Red Crescent Societies). 2014. https://www.ifrc.org/en/what-we-do/disaster-management/about-disasters/what-is-a-disaster/what-is-vulnerability/ (downloaded August 2014).

Iuchi, K., and A.M. Esnard. 2008. Earthquake impact mitigation in poor urban areas: The case of metropolitan Manila. *Journal of Disaster Prevention and Management*, 17(4), 454–469.

King, D. 2006. Planning for hazard resilient communities. In D. Paton and D. Johnston (eds.), *Disaster Resilience: An Integrated Approach* (pp. 289–299). Springfield, IL: Charles C. Thomas Publisher.

Laska, S., and Morrow, B. 2007. Social vulnerabilities and Hurricane Katrina: An unnatural disaster in New Orleans. *Marine Technology Society Journal*, 40(4), 16–26.

Miles, S.B., and Chang, S.E. 2006. Modeling community recovery from earthquakes. *Earthquake Spectra*, 22(2), 439–458.

Mileti, D. 1999. *Disasters by Design: A Reassessment of Natural Hazards in the United States*. Washington, DC: Joseph Henry Press.

Norris, F.H., Stevens, S.P., Pfefferbaum, B., Wyche, K.F., and Pfefferbaum, R.L. 2008. Community resilience as a metaphor, theory, set of capacities and strategy for disaster readiness. *American Journal of Community Psychology*, 41, 127–150.

NRC (National Research Council). 1994. *Practical Lessons from the Loma Prieta Earthquake*. Washington, DC: National Academies Press.

NRC (National Research Council). 2006. *Facing Hazards and Disasters: Understanding Human Dimensions*. Washington, DC: National Academies Press.

NRC (National Research Council). 2011. *Building Community Disaster Resilience through Private-Public Collaboration*. Washington, DC: Geographical Science Committee, National Research Council, National Academies Press.

OCHA (Office for the Coordination of Humanitarian Affairs). 2013. Haiti humanitarian action plan 2013. New York: United Nations.

Oliver-Smith, A. (ed.). 2009. *Development and Dispossession: The Crisis of Forced Displacement and Resettlement*. Santa Fe, NM: SAR Press.

Ordinario, C. 2012. Over 80% of PH economy prone to natural disasters. November 13. www.rappler.com/business/16058-more-than-80-of-the-philippine-economy-prone-to-naural-disasters-adb.

Paton, D. 2006. Disaster resilience: Building capacity to co-exist with natural hazards and their consequences (pp. 3–16). In D. Paton and D. Johnston (eds.), *Disaster Resilience: An Integrated Approach*. Springfield, IL: Charles C. Thomas Publisher.

Paton, D., and Johnston, D. 2006. Identifying the characteristics of a disaster resilient society (pp. 11–18). In D. Paton and D. Johnston (eds.), *Disaster Resilience: An Integrated Approach*. Springfield, IL: Charles C. Thomas Publisher.

Phillips, B.D., and Fordham, M. 2009. Introduction. In B.D. Phillips, D.S.K. Thomas, A. Fothergill, and L. Blinn-Pike (eds.), *Social Vulnerability to Disasters*. Boca Raton, FL: CRC Press.

Reiss, C.L. 2011. Enabling long-term recovery from disaster. *Periscope*, 14(2), 1.

Rivera, J.D., and Miller, D.S. (eds.). 2010. *How Ethnically Marginalized Americans Cope with Catastrophic Disasters: Studies in Suffering and Resiliency*. Lewiston, NY: Edwin Mellen Press.

Sanderson, D. 2000. Cities, disasters and livelihoods. *Environment and Urbanization*, 12(2), 93–102.

Sapat, A., and Esnard, A.M. 2013. Disaster planning and policies. Course prepared for FEMA Higher Education Program. Emmitsburg, MD: Emergency Management Institute.

Sapat, A., Mitchell, C.M., Li, Y., and Esnard, A.M. 2011. Policy learning: Katrina, Ike and post-disaster housing. *International Journal of Mass Emergency and Disasters*, 29(1), 26–56.

Scandlyn, J., Thomas, D.S.K., and Brett, J. 2013. Theoretical framing of worldviews, values, and structural dimensions of disasters. In D.S.K. Thomas, B. Phillips, W.E. Lovekamp, and A. Fothergill (eds.), *Social Vulnerability to Disasters* (pp. 33–56). 2nd ed. Boca Raton, FL: CRC Press.

Schwirtz, M. 2012. Housing nightmare looms in wake of storm. *New York Times*, November4.www.nytimes.com/2012/11/05/nyregion/freezing-temperatures-complicate-storm-recovey.html?ref = hurricanesandtropicals.

Sherrieb, K., Norris, F.H., and Galea, S. 2010. Measuring capacities for community resilience. *Social Indicators Research*, 99, 227–247.

Siolda, J. 2014. Razing San Francisco: The 1906 disaster as a natural experiment in urban redevelopment. Working paper. Waterville, ME: Economics Department, Colby College. https://jamessiodla.files.wordpress.com/2012/08/razing_siodla1.pdf (downloaded August 7, 2014).

Smit, B., and Wandel, J. 2006. Adaption, adaptive capacity and vulnerability. *Global Environmental Change*, 16, 282–292.

Smith, K. 1992. *Environmental Hazards: Assessing Risk and Reducing Disaster.* London: Routledge.

Spence, P.R., Eith, C., and Lachlan, A. 2010. Vulnerability of the underclass: Why the most vulnerable are always the last to know. In J.D. Rivera and D.S. Miller (eds.), *How Ethnically Marginalized Americans Cope with Catastrophic Disasters: Studies in Suffering and Resiliency* (pp. 317–335). Lewiston, NY: Edwin Mellen Press.

Tierney, K.J. 2007. Businesses and disasters: Vulnerability, impacts and recovery. In H. Rodriguez, E.L. Quarantelli, and R.R. Dynes (eds.), *Handbook of Disaster Research* (pp. 275–296). New York: Springer.

Trice, B. 2006. Urban management challenges in mega-cities: A survey of catastrophic events in the developing and developed world. *Urban Action: A Journal of Urban Affairs*, 69–79.

UNISDR (United Nations International Strategy for Disaster Reduction). 2009. *Global Assessment Report on Disaster Risk Reduction*. Geneva: United Nations.

United Nations. 2005. Press briefing by special rapporteur on right to adequate housing. www.un.org/News/briefings/docs/2005/kotharibrf050511.doc.htm.

Wisner, B. 2003. Disaster risk reduction in megacities: Making the most of human and social capital. In A. Kreimer, M. Arnold, and A. Carlin (eds.), *Building Safer Cities: The Future of Disaster Risk*. Washington, DC: World Bank.

Section IV

Planning and Path Dependence

10

The Historical Aspect of Social Vulnerability in the Lower Ninth Ward*

Nada Toueir

Contents

* This chapter is part of a doctoral research project, and some observations and statements are based on the researcher's field notes and analyses, newspaper articles, and census data. Multiple trips to New Orleans have been conducted, and the bulk of the data were collected during the summer of 2012 (May through August 2012). A special thank you to the Observatoire Ivanhoé Cambridge du Développement Urbain et Immobilier for their generous contribution (http://observatoire-ivanhoe-cambridge.umontreal.ca).

INTRODUCTION

Since Hurricanes Katrina and Rita struck in 2005, New Orleans and, more specifically, its most devastated neighborhood, the Lower Ninth Ward (LNW), have been the topics of many publications and research projects. This neighborhood has gained a lot of attention since these events took place, some of it positive and some negative. The LNW was the hardest hit by the hurricanes; it served as the "poster child" and turned into a tourist destination for middle- and upper-class Americans, mainly due to the very slow rebuilding process there. However, it also attracted people and organizations that were genuinely interested in rebuilding a better place for the residents who suffered and lost the most. It is a neighborhood with a unique trait and identity, but also with multiple contradictions and controversies, which makes it an even more interesting case to investigate. New Orleans is a neighborhood-oriented city, where each neighborhood identifies itself separately and where residents identify with their individual neighborhood. This became an issue in the rebuilding process after the city was hit by Hurricane Katrina, especially since 9 years after the calamity, the city and particularly the LNW are still suffering from the effects of the hurricane: blight, vacant properties, smaller population, uneven rebuilding, and much more (WhoData, 2009).*

This chapter does not focus on the events of August 29, 2005, but rather on the events between the creation of the neighborhood and that day in order to better understand the past and ongoing struggles and obstacles faced by the neighborhood and its residents that led to a partially rebuilt neighborhood several years after the hurricane. Why is the rebuilding process so slow in the LNW 9 years after Katrina? By looking into the historical evolution of the neighborhood, we realize that the hurricane was merely the triggering point that uncovered the real vulnerability rooted in the historical buildup of the city and the neighborhood.

This chapter intends to emphasize the complexity of the concept of vulnerability, especially social vulnerability, and look at it from a holistic and historical perspective, as most methods are not cumulative and seem to miss some aspect, rendering them only partially reliable in evaluating

* This can be deduced from simply following the daily news pertaining to New Orleans and reading the local newspapers, such as the *Times-Picayune* and Nola.com at http://www.nola.com/.

the vulnerability of a community. The focus on social vulnerability emanates from the fact that people die when disasters hit, mainly because they inhabit the built environment. Indeed, it is the combination of people living in cities that makes disasters so catastrophic. Also, the configurations of cities change in step with people's needs and thus evolve over time. These changes and the decisions leading to them can have either a positive or a negative impact on the built environment.

This chapter's contribution to disaster literature is to provide a better and more cumulative understanding of the historical evolution of a city and its complex constituent elements.

As researchers, we begin to fathom the consequences of this evolution on the present state of cities and learn from history in order to prevent the same mistakes from being repeated; this will allow us to propose better solutions that contribute to the prosperity of humans, rather than their deaths. According to Oliver-Smith and Hoffman (2002), "a disaster becomes unavoidable in the context of a historically produced pattern of 'vulnerability,' evidenced in the location, infrastructure, sociopolitical organization, production and distribution systems, and ideology of a society" (p. 3).

The chapter is divided into four sections. First, an overview of the concepts of vulnerability and social vulnerability is presented, with a focus on disasters and how they became a social construction, particularly in the case of Hurricane Katrina. Second, the focus is on the historical evolution of the Lower Ninth Ward, while highlighting the main events that contributed to its present-day configuration and state before discussing the different realities of the neighborhood: the social and cultural, the physical and environmental, the economic and political, and the differences and disparities existing within it. Third, how the historical evolution played an important role in making the Lower Ninth Ward a socially vulnerable neighborhood is demonstrated. Finally, the conclusion takes a look at the future and the importance of cultural identity in the neighborhood. Thus, the chapter aims to highlight the importance of analyzing the historical evolution of any area in order to understand its vulnerability, and more specifically its social vulnerability, while using the Lower Ninth Ward as a case study.

SOCIAL VULNERABILITY AND DISASTERS

From Vulnerability to Social Vulnerability

It is important to provide a quick overview of these definitions in order to understand the relevance of social vulnerability. Indeed, this concept helps in understanding the elements that contribute to a society's or city's destruction after any kind of disaster strikes. It highlights weak components at the physical, social, economic, political, and environmental levels in an effort to clarify and provide answers to why things happened the way they did. This literature review will present the concepts of vulnerability and social vulnerability, the tools used to evaluate and measure them, and the shortcomings of the latter due to the lack of consideration of time. Furthermore, the vulnerability literature provides multiple perspectives on how the concept is perceived, often leading to discrepancies in its meaning. These perspectives about vulnerability are summarized as follows: (1) vulnerability as a preexisting condition, (2) vulnerability as a response, and (3) vulnerability as a hazard of place (Weichselgartner, 2001). None of them concentrate on the importance of the historical evolution of the city, as history has a tendency to highlight the elements that contribute to the vulnerability of the study area. Vulnerability as a concept has elements of time that should also be considered when discussing the effects of social vulnerability. This chapter argues that vulnerability is constructed over time. The element of time has an important place in the discussion of vulnerability, as decisions play an important role in determining the vulnerability of a place or community. They contribute to either increasing or decreasing the risks involved in disasters. After all, these decisions "which have to be taken at the present, are neither correct nor incorrect today, but they can have good or bad consequences tomorrow" (Weichselgartner, 2001, p. 86).

Many researchers focus on the concept of vulnerability in social sciences and disaster research. It is beyond the scope of this chapter to define vulnerability, given that many have already defined it (Manyena, 2006; Weichselgartner, 2001; Thomas et al., 2012).

In addition to the way vulnerability is perceived, there are many definitions of the term, which has continuously evolved over time. In fact, many authors provided tables summarizing this (Weichselgartner, 2001; Manyena, 2006; Thomas et al., 2012); the first definition of the concept dates back to the early 1980s,* but the most cited and used definition of vulnerability is proposed by Wisner et al. (2004) in their book *At Risk*:

* "Vulnerability, therefore, will refer to the threat to which a community is exposed taking into account not only the properties of the chemical agents involved, but also the ecological situation of the community and the general state of emergency preparedness at any given point in time" (Gabor and Griffith, 1980, p. 2). "Vulnerability is the degree to which a system, or part of a system may react adversely to the occurrence of a hazardous event. The degree and quality of that adverse reaction are partly conditioned by the system's resilience, the measure of a system's, or part of a system's, capacity to absorb and recover from the occurrence of a hazardous event" (Timmerman, 1981, p. 21).

> By "vulnerability" we mean the characteristics of a person or group in terms of their capacity to anticipate, cope with, resist, and recover from the impact of a natural hazard. It involves a combination of factors that determine the degree to which someone's life and livelihood are put at risk by a discrete and identifiable event in nature or in society. Some groups in society are more prone than others to damage, loss and suffering in the context of differing hazards. Key characteristics of these variations of impact include class, caste, ethnicity, gender, disability, age, or seniority. (Wisner et al., 2004, p. 11)

According to the same authors, in order to evaluate and understand risk, "the social production of vulnerability" has to be considered as equally important as the natural hazard of interest, and they argue that "the risk of disaster is a compound function of the natural hazard and the number of people, characterized by their varying degrees of vulnerability to that specific hazard, who occupy the space and time of exposure to the hazard event" (Wisner et al., 2004, p. 49). The authors also proposed the following equation:

$$\text{Risk} = \text{Hazard} \times \text{Vulnerability}$$

As for social vulnerability, the definition that is used and cited the most is "social vulnerability is partially the product of social inequalities—those social factors that influence or shape the susceptibility of various groups to harm and that also govern their ability to respond" (Cutter et al., 2003, p. 243). These definitions will be used, as it is beyond the scope of the chapter to propose a new way of looking at these terms.

A lot of attention was given to assessing and evaluating the vulnerability of communities, neighborhoods, and cities. Although multiple methods exist, only two will be presented here: the Social Vulnerability Index (Cutter et al., 2003) and mapping social vulnerability (Thomas et al., 2012). Cutter et al. (2003) propose a measuring tool known as the Social Vulnerability Index (SoVI) to assess social vulnerability by determining a set of indicators that are indicative of the chosen study area. Their model is based on the premise that "social vulnerability is a multidimensional concept that helps to identify those characteristics and experiences of communities (and individuals) that enable them to respond to and recover from environmental hazards" (Cutter et al., 2003, p. 257). According to the authors, in order to determine the indicators used to calculate the SoVI, they must determine the factors affecting social vulnerability. These factors are cumulative and focus directly on the elements affecting a community at the social level. For Cutter et al., who compiled the work of other authors, these factors are "lack of access to resources (including information, knowledge, and technology); limited access to political power and representation; social capital, including social networks and connections; beliefs and customs; building stick and age; frail and physically limited individuals; and type and density of infrastructure and lifelines" (Cutter et al., 2003, p. 245). Even though they agree that the SoVI method seems to be a relevant tool to assess the vulnerability of an area and is useful as a comparative measure, they conclude that it is far from perfect and would need refinements due to the lack of information regarding previous disasters (Cutter et al., 2003). Indeed, to understand the social vulnerability of any urban community, measuring it can be a preliminary indicator of the state of the community, but not a cumulative one.

Others propose mapping vulnerability (Thomas et al., 2012) by dividing them into social vulnerability and territorial vulnerability. This method also relies on a series of relevant indicators that are computed and measured in order to portray the vulnerability of the studied area. This method is useful for city planners, decision makers, and citizens in understanding and visualizing the vulnerability of their area and is based on an iterative design for collecting and processing information. It relies not only on the indicators that need to be measured, but also on the different actors to weigh the indicators and thus present more accurate results. This method uses current information pertaining to the area of study and highlights the different elements contributing to its vulnerability. It involves some historical assessment (history of previous inundations and hydrology) but no in-depth evaluation of the decisions that affected the study area.

These two methods rely on numbers and maps, which are a more quantitative way to reach their conclusions. But when urban communities are involved, a more qualitative and historical approach should be used, one that is clearly lacking in the examples stated above. People's lives and their experiences cannot be summed up in numbers and maps alone. A historic look at the evolution of a city and the community and the reasons for their particular evolution before a calamity is crucial to understanding their combined vulnerabilities. Therefore, assessing the vulnerability of an area is a multilayered process that starts with a historical study of the area, followed by the combination of quantitative and qualitative data for a better assessment of the current conditions before and after the calamity. As it is beyond the scope of this chapter to tackle all of these elements, the aim is to focus on the historical aspect of social vulnerability and how it can assist decision makers and researchers in understanding the reasons behind the consequences of the disaster at issue and ways to improve the rebuilding process for a better and safer future. Ultimately, "in modern societies decisions are highly complex, interdependent, and interactive. They may either create increased risks, or they may reduce risks to potential disasters" (Weichselgartner, 2001, p. 86).

Natural Disasters: A Social Construction

> Eminently social, disasters are worked out in complex interactions and discourses in which the needs and interests of many involved individuals, groups, and organizations are articulated and negotiated over the often extended duration of the entire phenomenon. (Oliver-Smith and Hoffman, 2002, p. 12)

Disasters are not just events that take place and leave no trace behind them; the twentieth and twenty-first centuries have witnessed catastrophes that marked the history of the human race: earthquakes, volcanoes, hurricanes, and tsunamis. There will be no mention of wars here, as they are strictly human-induced disasters and the focus of this chapter is on natural disasters, whose consequences are far from natural, even though they are naturally induced at the source. These events turn into catastrophes when they hit a populated area that was built by humans. Disasters "take place through the conjuncture of two factors: a human population and a potentially destructive agent that is part of a total ecological system, including all natural, modified, and constructed features. Both of these elements are embedded in natural and social systems that unfold as processes over time" (Oliver-Smith and Hoffman, 2002, p. 3). Nevertheless, disasters are complex events that consist of multiple intertwined socially constructed facts (Oliver-Smith, 2002). Since disasters have consequences that are socially constructed, and cities are complex social, economic, political, and physical constructions, disasters become very complex events. These calamities cause the destruction/disruption of a complex system, the urban fabric, making disasters complex events with socially constructed consequences. In this chapter and according to the social constructionist thought, "a social construct is an idea or notion that appears to be natural and obvious to people who accept it but may or may not represent reality, so it remains largely an invention or artifice of a given society" (Social Constructs, 2008).

Since natural disasters are social constructions, the focus here will be on Hurricane Katrina. Yet, in the case of pre-Katrina New Orleans, the Lower Ninth Ward was already struggling with existing socially constructed problems like race, class, education, poverty, and social injustice, which contributed to its situation. As Erikson (2010) states:

> Katrina can be best understood as a collision between a natural force and what turned out to be strangely vulnerable social order. The true contours of Katrina cannot be learned from studying the intensity of its winds or the fury of its surges. They can be learned only from studying the other party to that collision—the ways in which humankind created the physical and social landscape the storm landed on and the ways in which persons and institutions responded to that crisis. That way of telling the story is a sociological one. (p. xx)

Hurricane Katrina had many faces and played multiple roles as a disaster: it was the costliest disaster in U.S. history, crippled an entire city for months, and caused the displacement of millions of people (Campanella, 2008). Most importantly, it exposed the social injustice and racial division not only in New Orleans, but also in the whole country. "What no social movement, no political party and certainly no sociological analysis, no matter how well grounded and brilliantly written (if such things existed!), would have been able to achieve what happened within a few days: America and the world were confronted by the repressed *other* America, the largely racialized face of poverty" (Beck, 2006, pp. 338–339).

Both the city and the neighborhood suffered from racial divides and tensions, which were highlighted even more after 2005; Gotham (2014) argues that the literature does not distinguish between the different aspects of racism, whether it is racial prejudice, discrimination, or institutional racism. For the author, "as a socially constructed and politically contested term, race is a historically changing concept that expresses a complexity of social meanings that are given concrete expression by the specific social relations and historical context in which they are embedded" (Gotham, 2014, p. 777). In the wake of Katrina, the combination of the city's location, the topography, the levees, racism, and the decisions made over the years resulted in a socially, politically, and economically constructed city, thereby explaining how Katrina became a socially constructed event that was triggered by nature, but with human-induced consequences.

Although the numbers show that the Lower Ninth Ward was a poor neighborhood prior to Katrina,* and remained one after the hurricane, it

* The Data Center, "Lower Ninth Ward Statistical Area," updated March 28, 2014, http://www.datacenterresearch.org/data-resources/neighborhood-data/district-8/Lower-Ninth-Ward/, and "Holy Cross Statistical Area," updated March 28, 2014, http://www.datacenterresearch.org/data-resources/neighborhood-data/district-8/Holy-Cross/.

was a very culturally rich community that relied on traditions, social networks, and a specific cultural identity that has been constructed over the years (Landphair, 2007). Without any financial and established resources, it becomes a challenge to overcome adversity because "the social production of *wealth* is systematically accompanied by the social production of *risks*" (Beck, 2013, p. 19). Risk is directly linked to class and wealth in that poor people are more at risk in the areas where they live, and they have fewer resources to rebuild, while the wealthy have the means to buy their safety and rebuild faster. Hence, "this 'law' of the class-specific distribution of risks and thus of the intensification of class antagonisms through the concentration of risks among the poor and the weak was valid for a long time and still applies today to some central dimensions of risk" (Beck, 2013, p. 35).

THE HISTORY AND THE DIFFERENT REALITIES OF THE LOWER NINTH WARD

The second section of the chapter sheds light on the historical evolution of the LNW and highlights the different realities that led to a very specific set of conditions before Hurricane Katrina arrived. Here, *reality* is referred to as constructed and defined "as a quality appertaining to phenomena that we recognize as having a being independent of our own volition," while *knowledge* is defined "as the certainty that phenomena are real and that they possess specific characteristics" (Berger and Luckmann, 1991, p. 1). The terms of interest are *reality* and *knowledge*. They have two contradicting roles, as they are parts of everyday language but have a very long history of philosophical scrutiny; therefore, what is known and what is real can be easily confused with the complexity that comes with them.

In the case of the LNW, the realities presented below were in fact a series of events that were taken for granted and clearly historically and socially constructed, and that explains the complexity of the situation. The existing knowledge that accumulated over decades of living and coping with disasters seemed to be all but missing when Hurricane Katrina hit. Whether this is due to the complete reliance on an unpreoccupied government, the intentional forgetfulness of past events, or the residents' lack of access to information, what happened to the city of New Orleans during and post-Katrina cannot repeat itself, and lessons should be learned from this devastating calamity. Therefore, it is imperative that researchers try to understand the facts that led to such a devastating state, as knowledge should be built on an understanding of the different processes that make reality a complex process because it is a social construction. "The need for 'sociology of knowledge' is thus already given with the observable differences between societies in terms of what is taken for granted as 'knowledge' in them. . . . In other words, a 'sociology of knowledge' will have to deal not only with the empirical variety of 'knowledge' in human societies, but also with the processes by which *any* body of 'knowledge' comes to be socially established *as* 'reality'" (Berger and Luckmann, 1991, p. 3). Consequently, by portraying the important historical facts and the different realities (social and cultural, physical and environmental, and economic and political) that shaped the LNW, this chapter will prove that these not only highlighted the complexity of the situation, but also were socially constructed and taken for granted when Katrina hit.

The Historical Evolution of the Lower Ninth Ward

The city was established in the early 1700s and grew and expanded gradually over time with major historical events: the Civil War, WWI, and WWII. It was founded by the French (early 1700s), occupied by the Spanish (in the 1760s), and then returned to the French (in 1800) until President Jefferson finally bought it from the French in 1803, thereby making it an American city (Campanella, 2008). The French and Spanish influenced greatly the architectural evolution of the city; the Creole, the Irish, the Vietnamese, and all the other immigrants influenced the food culture; and the city was home to jazz music and to a multiethnic population with a reputation that was recognized all over the nation and across oceans (Campanella, 2006). The reasons for building New Orleans were strictly economical and practical in nature. Its current site provided the optimal location for ships coming from Europe; in fact, the Port of New Orleans became the main source of income for the city, which attracted many people to settle there (Campanella, 2006). The first settlers could only build on the land along the Mississippi River, as it was the only land available above water, while the rest of the surrounding area was full of swamps and marshes. With technological advances, levees were built and the wetlands were drained so that the city could expand and the epidemics spread by mosquitoes could be controlled. The levees made it possible to develop land that was below sea level, which generated more revenue for the government and more affordable housing for the working class. Due to its incongruous topography, New Orleans is unevenly exposed to floods, as its neighborhoods are located at different elevations (Campanella, 2006). It also suffers from social, economic, and spatial inequalities that differ greatly from neighborhood to neighborhood.

The Lower Ninth Ward is one of the most recently developed neighborhoods in New Orleans. To better understand the historical formation of the area, it is imperative to take a look at the evolution of the city over the course of time, since it was the economic expansion of the city that led to the urbanization of the present-day Lower Ninth Ward. The major historical events that changed or transformed the neighborhood are summarized chronologically in Table 10.1.

Social and Cultural Realities of the Neighborhood

The present-day Lower Ninth Ward was mainly farmland before the 1840s; the area close to the Mississippi was above sea level and marshes were further inland, toward the north. In order to house its growing population, the best location for new development was downriver from the rich neighborhoods and downtown, the Ninth Ward (which now combines the Upper Ninth and the Lower Ninth Wards). Being a port city, New Orleans was a landing hub for newcomers and immigrants. In the 1850s, the city was one of the most culturally and ethnically diversified American cities and had the lowest percentage of locally born people (as low as 26%); the majority of the population was foreigners, specifically from countries such as Ireland, Italy, Haiti, and Vietnam, and all of them migrated and settled down in New Orleans, in the newly developed neighborhoods, such as the Ninth Ward (Campanella, 2006).

Therefore, New Orleans has always been known as or labeled an "ethnic gumbo" for its rich ethnic diversity; it was the home of people coming from different parts of the world, and many nineteenth-century authors described it as a heterogeneous ethnic city. Due mainly to the Civil War, New Orleans gradually started losing its reputation as America's most multicultural city in the late nineteenth century (Campanella, 2006; Falk et al., 2006). Over the years, the city also attracted a lot of freed African Americans; it was mainly the opportunity to work and the possibility of owning their own homes that contributed to moving there. It was a racially and ethnically mixed city with a working class that consisted of anyone who was willing to work hard for a living. In 1852, the Ninth Ward was created and included the present-day Upper and Lower Ninth Wards (Table 10.1 and Figure 10.1).

In the early and mid-1960s, when segregation ended and Hurricane Betsy hit, the LNW lost many of its residents but attracted many more. Its

Table 10.1 Important Dates That Led to the Transformation of the Lower Ninth Ward

Period/Date	Important Event(s)
Before 1840	The Lower Ninth Ward was predominantly rural and part of the Third Municipality (it included the Bywater area and the present-day LNW; refer to Figure 10.1).
1840	The first urban settlements emerged in the neighborhood.
1852	The city was divided into wards and the Ninth Ward was officially created.
1857	The first church, St. Maurice Catholic Church, was built, indicating that a large enough group of people lived in the neighborhood.
1910s	The first important change in the neighborhood was proposed: the Industrial Canal. The total population reached a high of 5,500 (1.6% of the total population) and consisted mainly of working-class, locally born and raised residents.
1923	The Industrial Canal was officially opened, which automatically created the Lower Ninth Ward (because it is located downriver).
1940s	After World War II, the population rose to 11,556 residents.
1960s	Whites left the LNW for St. Bernard Parish in reaction to school integration, which transformed the area into a predominantly African American neighborhood. The Mississippi River Gulf Outlet (MRGO), which later caused an environmental disaster, was initiated. The population fell sharply from more than 33,000 to 19,500 residents.
1965	Hurricane Betsy hit and the northern part of the LNW was completely flooded and devastated.
2000	What was a racially mixed LNW turned into a predominantly African American neighborhood (more than 95% of the population was black) with the highest home ownership rate in the city.
2005	On August 29, Hurricane Katrina hit New Orleans.
Post-Katrina	The LNW became a fragmented neighborhood that is still struggling to rebuild.

[a] All the facts in the table are taken from the book *Bienville's Dilemma: A Historical Geography of New Orleans* by Campanella (2008).

Source: Inspired by Campanella, R., *Bienville's Dilemma: A Historical Geography of New Orleans*, Center for Louisiana Studies, University of Louisiana at Lafayette, 2008. Compiled by the author.

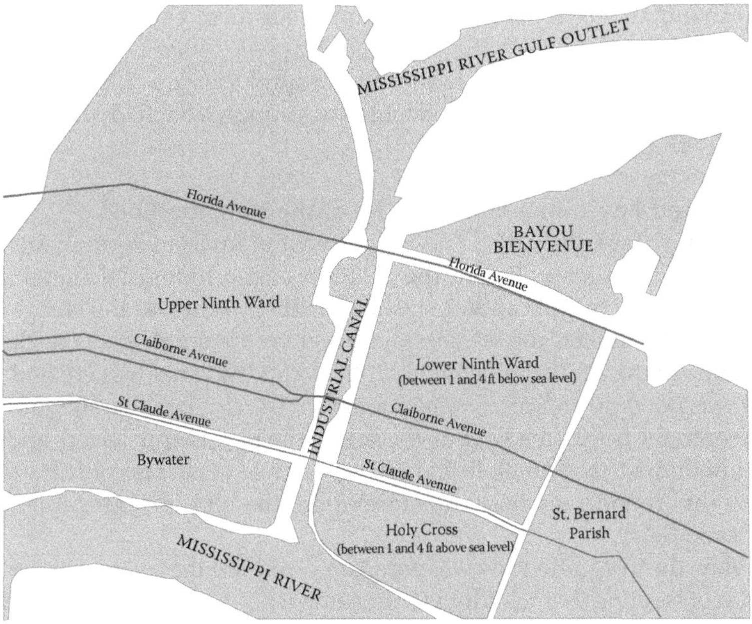

Figure 10.1 Map of the Lower Ninth Ward showing the main divisions within the neighborhood, the physical exclusion from the rest of the city, and the water surrounding the neighborhood from the north, east, and south.

population decreased dramatically when many white residents left, but they were replaced by family relatives of existing residents, new residents who wanted to move closer to the city and away from rural lands, and the residents of Fazendeville in particular, a very small community just downriver from the LNW (Chapman, 2004; Campanella, 2008). At that time, the government, which turned the area into a national park, bought out the residents of Fazendeville for a modest $6,000 to each family. So they were forced to move out of their community. The area was home to free people of color; it was a 100-year-old community of residents with very close bonds (Chapman, 2004). Not able to afford living in St. Bernard Parish, the majority of Fazendeville residents were forced to relocate and settle in a more affordable area that had lost the majority of its white residents and was more likely to accept newcomers, especially African Americans: the Lower Ninth Ward. The residents of Fazendeville settled there and moved their church, the Battleground Baptist Church, into the neighborhood as well (Chapman, 2004).* The combination of all these events explains the local culture of the LNW: the strong-knit aspect of the community, the reason why the majority of a family live around the same block and look after each other, and why the residents are strongly attached and identify greatly with their neighborhood.

Physical and Environmental Realities of the Neighborhood

As shown in Table 10.1, the LNW underwent many transformations that led to its current state. Before the creation of the Industrial Canal, and even though the neighborhood was physically attached to the city, it was sandwiched between the Mississippi River to the south and the Bayou Bienvenue to the north (Oliver et al., 2014). After the canal was created, the LNW was not only physically separated from the city, but also surrounded by water to the north (the Bayou), the south (the Mississippi River), and the west (the Industrial Canal). In the 1960s, a couple of events contributed to the neighborhood's greatest transformation: the end of segregation and Hurricane Betsy.

When the integration law was enforced, white families were outraged by the idea of sending their children to the same schools as black families, even though they lived next door to each other (*New Orleans*, 2007). Furthermore, Hurricane Betsy caused major damage to the LNW, flooding the majority of the neighborhood, especially the northern section, which was inhabited by black families; however, the Holy Cross area, which was mainly inhabited by white families, was spared by the floodwaters (Campanella, 2008). As a result, by the end of 1965, the majority of white families that previously lived in the neighborhood had moved to St. Bernard Parish (east of the LNW) and did not want anything to do with their old neighborhood (Campanella, 2008). By setting higher home prices, they made sure to keep the new neighborhood for white families like them, which created an invisible racial barrier that LNW residents knew not to cross.

Topographically, more than half the neighborhood lies below sea level due to the continuous pumping of water and is subsiding, as sediments are not deposited because of the ongoing water extraction (Campanella, 2006; Colten, 2009). The levees protecting the city are regularly maintained, and although they protect the city from the surrounding waters, the city is inundated right away whenever there is a breach, as witnessed in 2005. Also, when the city is exposed to floods, these levees play a dual role: they protect

* Ron Chapman, "Fazendeville," last visited August 28, 2014, http://www.myneworleans.com/Louisiana-Life/Winter-2004/Fazendeville/.

the city from floods and keep the water away, but they also create a false sense of security among the residents, who blindly trust that they will protect them, their homes, and their neighborhoods (Burby, 2006).

At the environmental level, specific actions resulted in a major deterioration of the natural ecosystem. Among these actions were the numerous canals and oil wells constructed in the wetlands to look for oil, which had a strong negative impact on this fragile environment. Moreover, a channel was excavated to connect the Mississippi River to the Gulf of Mexico: the Mississippi River Gulf Outlet (MRGO). The project was proposed in the late 1950s and was completed in 1968 (Campanella, 2006; Freudenburg et al., 2009). During the construction period, Hurricane Betsy (1965) hit and the resulting floods were higher than anticipated; this is how the MRGO earned the nickname "the hurricane highway" (Freudenburg et al., 2009). The MRGO not only provided a highway for the seasonal hurricanes, but also greatly impacted the area's biodiversity and wetland environment. By connecting the Gulf of Mexico to Lake Pontchartrain, salt water found its way into the lake's freshwater, which affected the local fish and shrimp populations, as well as the forest of cypress trees that held the fragile soil together. As cypress trees do not thrive in salt water, they gradually died, exposing the city to stronger winds and contributing to the loss of land, which widened the canal over the years. The loss of land meant that the canal needed to be dredged on a regular basis in order to maintain its depth so that larger ships could pass through. However, this allowed more salt water to enter and kill more cypress trees. It was a vicious circle that came to an end when the city finally decided to close the MRGO in 2008 (Freudenburg et al., 2009).

The Economic and Political Realities of the Neighborhood

Early on, when the LNW was an extension of the Ninth Ward, it was easiest to build along the shores of the Mississippi River since this land was naturally higher than that located further north. As the city's population started to grow and the city attracted more and more immigrants and freed African Americans, space became limited, but housing all these newcomers in the older or wealthier neighborhoods was not considered an option. As a result, the Ninth Ward kept growing. Ultimately, the construction of the levees provided an opportunity to build more affordable residences, making it possible for the working class to own their homes. This phenomenon explains the high homeownership rates in the LNW prior to Hurricane Katrina (Campanella, 2008).

Also, the numerous decisions made over the years demonstrated a weak political strategy toward the LNW. The first of these was the construction of the Industrial Canal, which physically cut off the LNW from the rest of the city. Then the levees were constructed to expand residential developments, but this was done according to questionable building and safety standards (Lee, 2010). Finally, the creation of the MRGO's canals made it easy for hurricanes to enter the city. These decisions lead to one conclusion: New Orleans city management did not learn its lesson from Hurricane Betsy, as most of these decisions were made around that time (Campanella, 2008).

The Differences and Disparities within the Neighborhood

As discussed earlier, the LNW was built for the specific purpose of housing the working class and the increasing number of immigrants that resided in New Orleans. It was a practical and easy solution: drain the marshes to build affordable residential homes far from the central and rich areas but close enough to the city to be convenient to residents.

The natural levee along the Mississippi River protects the southern area of the LNW from other flooding, except when the river rises above the natural levee's limits. Over the years, the area between St. Claude and the river, known as Holy Cross (Figure 10.1), was spared from high waters. This asset allowed the area to be developed before the levees were built, and as it grew, it became home to newcomers and the wealthier families who came to work in the city. As the levees were built, the area north of St. Claude, which was mainly marshes and wetlands, was drained in order to house more working-class families (Figure 10.1).

The topography in the Lower Ninth Ward varied. The further homes were from the Mississippi River, the easier they flooded because they were on low-lying grounds. Over time, the neighborhood was sinking, with some areas to the north ending up 4 feet below sea level (Campanella, 2008). After the building of the Industrial Canal, the disparities between Holy Cross and the Lower Ninth Ward grew, even though they were both physically separated from the rest of the city. Each subneighborhood turned inward and took care of its own. Halfway through the century, the end of segregation and the arrival of Hurricane Betsy, just a few years apart, forced the white population and those who could afford it to leave the LNW, which turned it into a majority low-income African American neighborhood. Moreover, despite the numerous floods that occurred throughout the history of the Lower Ninth Ward, the Holy Cross area suffered less damage than its northern counterpart, which kept older homes in better condition; its homes were maintained and cared for because it was an area with slightly wealthier families. After Katrina, Holy Cross received the status of historic district from the city of New Orleans's Historic District Land Commission, and many of the homes that were not destroyed were renovated by the city according to specific standards (City of New Orleans Historic District Land Commission, 2013).* The cultural divide between the Holy Cross area and the Lower Ninth Ward was engraved into people's memories and identities, before and after Katrina.

A SOCIALLY VULNERABLE LOWER NINTH WARD

Hurricanes Katrina and Rita affected the whole southeast coast of the United States and especially New Orleans. However, each of its neighborhoods was affected differently depending on its specific configuration, topography, and socioeconomic status. Each community living in these neighborhoods also suffered the consequences differently, as "not all communities experience a disaster in the same way or to the same degree; each undergoes a catastrophe in the context of its own profile of vulnerability" (Oliver-Smith and Hoffman, 2002, p. 13). Even though 80% of the city was flooded and damaged, each of its neighborhoods rebuilt differently and at a different pace (Campanella, 2008). The LNW was the most devastated and destroyed in the city and struggles even now with the rebuilding process.

Although Lakeview was also one of the most heavily flooded neighborhoods in New Orleans, it did not suffer from the water surge that the LNW faced when the levees broke. The water level rose in Lakeview but did not leave its trace by displacing houses. Cars were not stacked on top of each other or on top of homes in Lakeview, and the majority of its residents belonged to the middle- and high-income class. The LNW residents had to wait months before they received authorization to come back to their property and evaluate the damage. The neighborhood was closed off to its own residents, whereas none of the other neighborhoods faced such obstacles and were able to start the rebuilding process much faster than the LNW.†

Why is it important to talk about social vulnerability in the Lower Ninth Ward? In the case of the LNW, the elements that made it culturally

* The City of New Orleans's Historic District Land Commission, "Historic District Maps and Location Information," updated November 6, 2013, http://www.nola.gov/hdlc/map/.

† This can be deduced from simply following the daily news pertaining to New Orleans and reading the local newspapers, such as the *Times-Picayune* and Nola.com at http://www.nola.com/.

unique and held it together as a neighborhood and community became the very elements that contributed to its vulnerability. Since its inception, and even when it was still part of the Ninth Ward, this neighborhood was vulnerable in different ways. The majority of people living in the Ninth Ward were either immigrants or poor. It had many nicknames as well: "the Creole faubourgs," the "old Third," the "dirty Third," the "poor Third," and paradoxically, the "glorious Third" (Campanella, 2008). The entire city looked down upon the neighborhood and its residents; it was never an area of economic, political, or social interest.

With improved technology, the city built bigger and stronger levees and was able to drain the surrounding marshes and wetlands in order to build more affordable housing. Just under half of the LNW's parcels are above sea level (the southern part that is closer to the Mississippi River), and more than half of its parcels are at or below sea level (the northern part, north of Claiborne Avenue). The blocks on higher grounds were the first to be developed and owned by slightly richer residents, which created tensions within the two subneighborhoods that are still present in the LNW today (Campanella, 2008).

As shown in the historical evolution earlier, this area was developed in the later stages of the city's expansion to house the working class and keep them away from, yet close enough to, uptown and downtown; this is due mainly to the fact that they were poor yet essential to the growth of the city. An overview of the major historical events that led to the formation and transformation of the LNW shows us that the combination of all these facts explains many of the consequences of Hurricane Katrina. At the social and cultural level, the affordability of the neighborhood combined with the many events mentioned earlier ended up attracting low-income African Americans to the LNW, which transformed it from a racially mixed and working-class neighborhood into a majority African American, low-income neighborhood.

The LNW has an emotional value only to its own residents; people had to "pass through" it to go from St. Bernard Parish to New Orleans and vice versa, but never had a reason or purpose to stop there. According to Campanella (2008), "the rear sections of the Lower Ninth Ward seemed like a world unto itself—cherished by its residents, avoided by everyone else." Because the Industrial Canal physically separated it from the rest of the city, LNW residents turned to each other for support and help. The canal made the residents feel unwanted and uncared for by the government, which brought them closer to each other. The local culture of the neighborhood was geared toward strong family values, church, music, and food. This proximity and the close-knit aspect of the community also provided a sense of comfort and stability to residents. It helped them form a unique cultural identity, which turned out to be a curse when destructive hurricanes like Katrina and Rita struck. By that time, entire families lived on or around the same block, and mothers and grandmothers looked after the family's children while parents went to work (Baum, 2009; Jackson, 2005). Since homes were handed down from generation to generation, the LNW was one of the neighborhoods with the highest home ownership rates in the city. However, after Katrina, the entire family structure was broken; when a family would go next door to seek help and assistance from relatives, their relatives were unable to help, as everyone faced the same struggles and was in dire need of help. Entire networks of friends and families were adversely affected by the same calamity, and no one could turn to their neighbors anymore. They had to face life and death situations all alone.

At the physical and environmental level, the decisions to build the levees, the canal, and the MRGO show the disinterest in integrating the LNW into the rest of the city, as "the federal levee failures induced by Hurricane Katrina and the preceding century of environmental deterioration altered utterly the destiny of the Lower Ninth Ward" (Campanella, 2008). Whether it was due to race or wealth or a combination of both, it seems that separating the neighborhood was a careless move that ultimately accentuated its weaknesses and prevented it from flourishing. Indeed, some numbers and facts show a different reality: "Once racially mixed, the neighborhood in 2000 was over 95% black. By no means was the Lower Ninth Ward the poorest or lowest-lying neighborhood of the city. It actually boasted a higher home-ownership rate than the city as a whole, and its lowest-lying areas (four feet below sea level) lay three to four feet above the *lowest* zones of Lakeview and Gentilly, and eight feet higher than the lowest spots in New Orleans East" (Campanella, 2008, p. 153). This explains how the vulnerability of this community was socially constructed, rendering it more fragile and susceptible to an event as destructive as Katrina.

Although the strong ties between the residents of the LNW created a sense of attachment and identity, their negative image of themselves contributed to the neighborhood's vulnerability. First, it was sociocultural exclusion, then physical and environmental exclusion, and finally economic exclusion. Disasters, specifically Katrina, revealed the reality behind the social structure of its fragile society and not only destroyed everything in their path, but also exposed the lack of cohesion and the weak areas of the LNW. It showed the closeness and kinship that existed among the neighbors, but it also displayed to them and the rest of the world the virtual lack of commitment and involvement of the local government. Oliver-Smith and Hoffman (2002) wrote, "The distribution of power within a society reveals itself not only in the differential vulnerability of groups, but in the allocation of resources in reconstruction as well. Disasters provide a unique view of a society's capacities for resistance and resilience in the face of disruption" (pp. 9–10). Even during post-Katrina reconstruction, the subdivisions within the neighborhood are still present. The residents still consider the two subneighborhoods (the LNW and Holy Cross) separate and resent one another for getting more attention from the city when there is only a street separating them (St. Claude Avenue).

As a result, the LNW seems to have fallen into a pattern that reflects its historical evolution. The sequence of events that created this pattern in the first place explains many of the reasons why this neighborhood suffered and endured as much as it did in 2005: "a society's pattern of vulnerability is a core element of a disaster. It conditions the behavior of individuals and organizations throughout the full unfolding of a disaster far more profoundly than will the physical force of the destructive agent" (Oliver-Smith and Hoffman, 2002, p. 3). Therefore, the reasons and implications for the elements of vulnerability affecting the LNW resulted from a complex nature–society relationship induced by a sociocultural construction.

CONCLUSION

Because it is surrounded by water, the city of New Orleans has always been prone to hurricanes and floods. Over the course of its history, hurricanes became part of the local culture. The city was first built along the safer shores of the Mississippi River located above the water level; however, as it is surrounded by the Mississippi River and various marshes, New Orleans's relationship to water is intricate and delicate. Water is the source of its success and strength, but also its vulnerability and weakness; this element of nature is at the root of the city's rich and diversified culture, but also its destruction and heartache. Residents coped and learned to live with floods over the centuries. In recurring disasters, local residents' experience, survival, and cultural identities are in direct correlation with the disaster itself. Disasters in general, but especially ones like Hurricanes Katrina and Rita, "unmask the nature of a society's social structure, including the ties and resilience of kinship and other alliances" (Oliver-Smith and Hoffman, 2002, pp. 9–10). In the case of the LNW, the residents faced different floods, but nothing as devastating as what happened in 2005. Many asked various questions: Why rebuild in a vulnerable area? Why not build higher? Why is the LNW not rebuilt yet? There are many answers to all these questions, and they all lie in understanding the neighborhood's historical evolution.

Lessons should be drawn from the LNW and Katrina in order to prevent such loss and devastation from occurring again. It is also very important to study the community at stake. The cultural identity and values of the LNW's residents are very specific and reflect the identity of the whole neighborhood because calamities and how societies deal with them over time "are potential indices of not only appropriate environmental adaptations, but ideological ones as well. These cultural adaptations include innovation and persistence in memory, cultural history, worldview, symbolism, social structural flexibility, religion, and the cautionary nature of folklore and folk tales" (Oliver-Smith and Hoffman, 2002, p. 9).

The context of this chapter illustrates that disasters are engraved in local culture and are part of the local identity. "There is much to be learned about cultural and societal modification from the calamities a people endure. Not only did societies undertake immediate adaptations after impact, but these can also set in motion forces with long-term implications for the evolution of each society" (Oliver-Smith and Hoffman, 2002, p. 9). The cultural identity and strong ties existing in the LNW played an important role in whether people returned to or abandoned their neighborhood. The residents who are now struggling with the post-Katrina efforts to resurrect their neighborhood are the same residents who have not and will never abandon the LNW (Landphair, 2007). But if hurricanes are such an important part of the local culture, then where was this "adaptation" when Katrina hit? Did their history of hurricanes contribute to their vulnerability or not?

The LNW is only one of the neighborhoods of New Orleans, and looking at its own vulnerability is one aspect of a much more complex situation. The LNW, despite all that it suffered and continues to suffer, is part of a larger entity: the city of New Orleans. Even though the factors affecting its vulnerability can be traced back to the inception of the neighborhood, the LNW is nevertheless part of an even more complex urban system: the city—a city that has a rich history behind its creation, its development, and its vulnerability. This chapter sheds light on the importance of considering the historical evolution of any neighborhood, community, or city in an attempt to understand its vulnerabilities in the face of calamities. By understanding the various subtleties of these vulnerabilities, researchers and decision makers (especially at the governmental level, local and national) can pinpoint the elements that need to be improved in order to avoid devastation on a scale like that experienced in New Orleans. Planners, researchers, and local officials should work hand in hand to propose short-term and long-term solutions to vulnerabilities that result from decisions made in the past and, when exposed by an agent as devastating as Katrina was to New Orleans, have devastating consequences. The culture of disaster research and disaster management should be looked at from a different angle: understand the vulnerabilities and how they were constructed over the course of history, and provide sustainable deep-rooted solutions to minimize the vulnerabilities and increase resistance; otherwise, the same devastation will take place over and over again.

REFERENCES

Baum, D. 2009. *Nine Lives: Death and Life in New Orleans*. New York: Random House LLC.

Beck, U. 2006. Living in the world risk society. *Economy and Society* 35(3):329–345.

Beck, U. 2013 [1992]. *Risk Society Towards a New Modernity*. London: Sage.

Berger, P. L., and T. Luckmann. 1991. *The Social Construction of Reality: A Treatise in the Sociology of Knowledge*. New York: Penguin.

Burby, R. J. 2006. Hurricane Katrina and the paradoxes of government disaster policy: Bringing about wise governmental decisions for hazardous areas. *Annals of the American Academy of Political and Social Science* 604(1):171–191. doi: 10.1177/0002716205284676.

Campanella, R. 2006. *Geographies of New Orleans: Urban Fabrics before the Storm*. Lafayette: Center for Louisiana Studies.

Campanella, R. 2008. *Bienville's Dilemma: A Historical Geography of New Orleans*. Lafayette: Center for Louisiana Studies, University of Louisiana at Lafayette.

Chapman, R. 2004. Fazendeville. http://www.myneworleans.com/Louisiana-Life/Winter-2004/Fazendeville/ (accessed August 28, 2014).

The City of New Orleans Historic District Land Commission. 2013. Historic district maps and location information. Last Modified November 6, 2013. http://www.nola.gov/hdlc/map/ (accessed August 30, 2013).

Colten, C. E. 2009. *Perilous Place, Powerful Storms: Hurricane Protection in Coastal Louisiana*. Jackson: University Press of Mississippi.

Cutter, S. L., B. J. Boruff, and W. L. Shirley. 2003. Social vulnerability to environmental hazards. *Social Science Quarterly* 84(2):242–261.

Erikson, K. T. 2010. Foreword. In *The Sociology of Katrina: Perspectives on a Modern Catastrophe*, ed. D. L. Brunsma, D. Overfelt, S. Picou, xxvii–xx. Lanham, MD: Rowman & Littlefield Publishers.

Falk, W. W., M. O. Hunt, and L. L. Hunt. 2006. Hurricane Katrina and New Orleanians' sense of place: Return and reconstitution or "gone with the wind"? *Du Bois Review* 3(1):115–128.

Freudenburg, W. R., R. Gramling, S. Laska, and K. T. Erikson. 2009. *Catastrophe in the Making: The Engineering of Katrina and the Disasters of Tomorrow.* Washington, DC: Island Press/Shearwater Books.

Gabor, T., and T. K. Griffith. 1980. The assessment of community vulnerability to acute hazardous materials incidents. *Journal of Hazardous Materials* 3(4):323–333.

Gotham, K. F. 2014. Racialization and rescaling: Post-Katrina rebuilding and the Louisiana Road Home program. *International Journal of Urban and Regional Research* 38(3):773–790.

Jackson, W. 2005. *What Would the World Be without Women? Stories from the Ninth Ward.* Neighborhood Story Project.

Landphair, J. 2007. The forgotten people of New Orleans: Community, vulnerability, and the Lower Ninth Ward. *Journal of American History* 94(3):837–845.

Lee, S. 2010. If God Is Willing and Da Creek Don't Rise. HBO documentary.

Manyena, S. B. 2006. The concept of resilience revisited. *Disasters* 30(4):433–450.

New Orleans. (2007). DVD. Directed by Steven Ives. Public Broadcasting Service.

Oliver, A., I. Thomas, and M. M. Thompson. 2014. Resilient and regenerative design in New Orleans: The case of the Make It Right project. *SAPIENS Surveys and Perspectives Integrating Environment and Society* 6.1.

Oliver-Smith, A. 2002. Theorizing disasters: Nature, power, and culture. In *Catastrophe and Culture: The Anthropology of Disaster,* ed. A. Oliver-Smith and S. M. Hoffman, 23–48. Santa Fe, NM: School of American Research Press.

Oliver-Smith, A., and S. M. Hoffman. 2002. Introduction: Why anthropologists should study disasters. In *Catastrophe and Culture: The Anthropology of Disaster,* ed. A. Oliver-Smith and S. M. Hoffman, 3–22. Santa Fe, NM: School of American Research Press.

Social constructs. 2008. *International Encyclopedia of the Social Sciences.* Encyclopedia .com. http://www.encyclopedia.com/doc/1G2-3045302485.html (accessed November 11, 2014).

Thomas, I., N. Bleau, P. Soto Abasolo, G. Desjardin-Dutil, M. Fuamba, and S. Kadi. 2012. Analyser la vulnérabilité sociétale et territoriale aux innondiations en milieu urbain dans le contexte des changements climatiques, en prenant comme cas d'étude la ville de Montréal. Rapport final pour Ouranos. *Ouranos,* p. 137.

Timmerman, P. 1981. *Vulnerability, Resilience and the Collapse of Society: A Review of Models and Possible Climatic Applications.* Environmental Monography. Toronto: Institute for Environmental Studies, University of Toronto.

Weichselgartner, J. 2001. Disaster mitigation: The concept of vulnerability revisited. *Disaster Prevention and Management* 10(2):85–95.

WhoData. 2009. WhoData.org. http://www.whodata.org/ (accessed August 30, 2014).

Wisner, B., P. M. Blaikie, T. Cannon, and I. Davis. 2004 [1994]. *At Risk: Natural Hazards, People's Vulnerability, and Disasters.* 2nd ed. London: Routledge.

11

Building Back Tremé

Using PPGIS to Evaluate Neighborhood Stability

Michelle M. Thompson, Brittany N. Arceneaux, and Grace Elizabeth Major

Contents

INTRODUCTION

When Hurricane Katrina struck the Gulf Coast on August 29, 2005, the world watched as New Orleans and the surrounding area suffered one of the most destructive natural disasters in recent history. In the months to follow, the media reeled with updates on the condition of the city, harrowing personal accounts, and the debate about how to proceed. Many tried to condemn New Orleans—for being too far below sea level, for having lost its relevance in the twentieth century, for its crime rates and poverty level—and argued to shut it down altogether. However, instead of abandoning the Crescent City, residents were overcome by the support of local, national, and international volunteers. Since it has been almost a decade since Hurricane Katrina, it is apparent that their efforts were not in vain.

Hurricane Katrina was not the first major storm to hit New Orleans, nor with certainty can it be determined to be the most destructive. The first recorded hurricane struck the nascent colonial settlement in 1722, destroying the existing structures on the riverbank and requiring a complete rebuild (Campanella, 2008, p. 22). The combination of factors that allow places such as New Orleans to regenerate is now identified by scholars as part of resilience. The type, measurement, and scale of factors are being studied and promoted for their benefits to communities as well as impact on built environments. In research provided, resilience and its relationship to two other urban phenomena, gentrification and neighborhood stabilization, will be examined through the lens of the Tremé neighborhood of New Orleans, Louisiana. Neighborhoods that are affected by either man-made or natural disasters require alternative methods to evaluate if, how, and when they have moved from *recovery* to *stabilization* or from *shock* to *resilience.* "A shock is defined as a sudden event that impacts on the vulnerability of a system and its components" (Roussey, 2013, p. 4). The information available to determine neighborhood stability postshock must include traditional and innovative approaches to integrate data.

The Success Measures neighborhood stability evaluation framework includes market strength in terms of tracking the visible physical conditions through field surveys, measuring neighborhood image and confidence through interviews of both residents and key outside informants, and gathering key baseline demographic and socioeconomic data (Neighborworks, 2011). "National studies, however, have usually relied on census tracts or other combinations of blocks as the best available proxy for a neighborhood" (Stewart, 1996, p. 39).

In order to measure the existence or change in gentrification, resilience, and neighborhood stabilization, similar qualitative and quantitative measures are utilized. Gentrification has varying definitions, but each includes ideals about municipal or community governance and revitalization. Some argue that this neighborhood change may lead to more stable communities that decimate social networks. "At its root, gentrification is the phenomenon of demographic succession (from poor or working-class to upper-income, from less-educated to more-educated, and from minority to white), driven by market forces in poor, minority neighborhoods" (Tufts University, 2009, p. 5). This neighborhood change uses comparative and longitudinal studies that are influenced by similar baseline indicators. Most of the indicators use public information that is typically linked to census data. Public participation geographic information systems (PPGIS) in particular have become a complement to these data sources. Having considered a number of models, the goal of this research follows GIS empirical analysis that employs volunteered geographic information (VGI) data and "easily obtainable variables from administrative sources that will [use] neighborhood quality-measurement techniques" (Rybarczyk and Mohapatra, 2013). VGI is "crowdsourced" geographic information provided by a wide range of participants with varying levels of education, knowledge, and skills (Global Facility for Disaster Reduction and Recovery, 2014). In the case of Tremé, property condition data were obtained from primary and secondary data sources. The primary data were collected by residents and volunteers using property condition survey standards of WhoData.org. Municipal secondary data remain limited and many times need conversion to a spatial format. For example, GIS analysts obtained the location and disposition of blighted properties summarized in the city of New Orleans Blightstatus (City of New Orleans Blightstatus, 2014) database. By combining the municipal and community data into a single data set for mapping and analysis, the Tremé project demonstrated VGI model concepts.

Through an evaluation of the literature, we suggest a model that supports the use of spatial data to aid in measuring neighborhood stabilization. Crowdsourced data are typically current and provide the "ground truth" of municipal data. VGI data are not typically available but can be considered a reliable indicator of neighborhood change. For the purpose of this study, data collection and analysis were standardized using Federal Geographic Data Center (FGDC) and WhoData.org data collection models.

Measurement of neighborhood changes in sociodemographic, housing, employment, and land use patterns in New Orleans has been in place since 1718. The neighborhood life cycle of Tremé suggests that the magnitude of gentrification has been significant and verifiable since the 1980s (Gladstone and Préau, 2008). Tremé has been impacted by the redefinition of neighborhood boundaries for political reasons (e.g., tourism) and affected by infrastructure design (e.g., construction of Interstate 10 starting in 1966). For the purpose of this chapter and to compare VGI with municipal data, the boundaries of the city planning commission will be used for analysis. The neighborhood boundaries and the community name have varied since its origin. The process of integrating VGI and municipal data is based upon the use of applied geographic information systems, specifically *public participation geographic information systems*. The quantitative measures used to identify the socioeconomic trends will include variables found in postdisaster studies of resilience and city change cycles defined as gentrification. This study will examine literature that describes neighborhood change models used to define and measure neighborhood stabilization. The next section will provide a history and summarize neighborhood trends in the Tremé neighborhood. The PPGIS model framework, using spatially integrated VGI with publicly accessible data, will be described. Scholars and planners in practice should consider the benefits of scalable data for neighborhood change measurement. The next section includes the case study of Tremé PPGIS mapping and analysis. Limiting conditions of the applied model will be followed by the conclusion.

DEFINING NEIGHBORHOOD STABILIZATION

Many of the variables used to evaluate neighborhood change, desirability, resilience, and stability are similar. The social and economic policies and planning initiatives were initially based on ethnic or race-based theories that have since proved faulty. The neighborhood life cycle was initially developed by the real estate industry, and then adopted by urban planners. This cycle suggested that there was an inevitable decline in neighborhood desirability when housing conditions and race changed from excellent to poor, from white to black. The real estate industry was initially supported by the Federal Housing Administration (FHA) and did not initially refute the risk factors that included using race as a basis for "redlining" to prevent funding for sale or rehabilitation in certain neighborhoods. Whether it was the five-stage Housing and Urban Development (HUD) neighborhood life cycle or the Real Estate Research Corporation's (RERC) life cycle model, the ability for ethnic minorities, in particular African Americans, to receive financing was limited due to perceived risk. The RERC life cycle model accommodated "racial infiltration" theories by encouraging planners to downgrade neighborhoods where African Americans lived (Metzger, 2000). Fair Housing and HUD funding increased opportunities to access affordable housing by establishing the first Neighborhood Housing Services (NHS) in 1968 (AllGov, 2014). NHS became Neighborworks America (also known as the Neighborhood Reinvestment Corporation) in 1978 to increase housing access for low- to moderate-income residents through homebuyer education, financial fitness, and funding to purchase homes in areas previously deemed marginal or blighted.

As HUD and the RERC moved away from using race-based measures to define neighborhood change, block-level data and economic shifts were used to establish the health of a community. The ability to reverse the adverse effects of disinvestment came in many forms. Banks were required to expand federally backed mortgages in previously redlined neighborhoods through the Community Reinvestment Act of 1977. The Financial Institutions Reform, Recovery, and Enforcement Act of 1989 (FIRREA) required that real estate appraisals exclude race as a factor in valuation. HUD created Community Development Block Grant (CDBG) funds in 1974 to expand affordable housing options through dedicated development opportunities. The purpose of measuring trends changed from focusing on the stages of decline to evaluating the ability of a community to remain resilient, foster growth, and maintain stability. Neighborhood stabilization can be defined as "the process of fostering market recovery by reversing destabilizing trends and rebuilding resident and homebuyer confidence" (Mallach, 2008, p. 5).

Measuring market recovery, revitalization, resilience, gentrification, and neighborhood stability is based upon many of the same variables: demographics, employment, housing trends, transportation access, education, infrastructure, crime, and social networks. How these variables are measured, integrated, and evaluated depends upon the data available, the level of geography, and data definitions and exclusions. Many of the studies focus on city or regional analyses, since household and parcel-level data are difficult to obtain or not available. The discussion of resilience in this context has to do with the environmental and psychosocial aspects that affect, at the neighborhood level, the ability for the individual, household, and neighborhood to recover. "Because of the multidimensional nature of resilience and its different component parts, a broad model of resilience has yet to be empirically tested at the community level. However there is consensus within the research community that resilience is a multifaceted concept, which includes social, economic, institutional, infrastructural, ecological, and community elements" (Cutter et al., 2010, p. 6).

The focus of this research is to use standard quantitative measures of evaluating neighborhood change by examining the case study of Tremé. "We recognize that exogenous factors such as federal policies and state regulations do exert powerful influences on resilience at the community level" (Cutter et al., 2008, p. 602). Externalities such as interstate development and urban renewal decimated the social and economic networks of this historic African American community. We will not focus on the social aspects but consider that "the relationship between homeownership and neighborhood stability is likely to be reciprocal. That is, living in a relatively stable neighborhood will further encourage participation in community organizations, local social interaction and attachment, property maintenance, neighborhood satisfaction, and positive expectations about the future of the neighborhood" (Rohe and Stewart, 1996, pp. 54–55). The issue of changes that result from processes such as job loss, recession, natural disaster, uneven redevelopment, increases in rent, value anticipation which leads to price inflation and lack of community cohesion, loss of leadership, fracturing of culture tied to the place can be measured using a "Disaster Resilience of Place (DROP) model" (Cutter et al., 2010, p. 5). The DROP model is based primarily on theoretical notions of resilience after measurement of vulnerability and risk. Cutter et al. suggest that

> the next step is to operationalize the model, develop a set of common indicators, and then test it in a real-world application. This necessitates additional research on resilience metrics. Such an application should provide sound measurements for assessing what makes some places more resilient in the face of natural disasters than others and would permit the comparison of community resilience over time and across space using the same set of measures. It should provide the guidance for implementing more sustainable practices that empower local communities to take their risks seriously, and at the same time provide guidance on the structural, economic, social, and environmental policy changes needed to enhance their own resilience. (Cutter et al., 2008, p. 604)

The Tremé research will apply the framework of the DROP model to a real world situation by

1. Identifying federal, regional, and local indicators that can be analyzed using common definitions and standards
2. Using variables that are measured at different levels of geography but within the same time frame
3. Creating a framework for capturing data before, during, and after a disaster (natural or man-made) event
4. Empowering residents to engage in collection, management, and evaluation of VGI data to engage in policies and plans that aid in measuring and supporting neighborhood change

While it may be another source of data that could be used to triangulate quantitative measures, we did not collect any new qualitative data.

The Katrina fatigue effect (which focuses on the reduction of compassion by those not affected directly by the storm) on the interview responses (Pezzullo, 2010) was a concern. We considered alternative ways to diffuse potential data reliability, validity, and error when conducting interviews of Tremé residents. There is significant variability of former to current residents based on demographics (age, gender, race, etc.), homeownership type (renter vs. homeowner), tenure, education, and employment type. Additionally, in the case of New Orleans in general, and the case study of Tremé in particular, there was a fatigue related to the incessant community meetings, surveys, and resident interviews. Survivor fatigue included the emotional, psychological, and sometimes physical effects that could skew responses from disaster experiences due to varying levels of trauma. To some extent, informal discussions about the pre- and post-disaster history of Tremé were used to inform the study. Given this, and understanding that the residents included in the study had limited tenure and demographics from longer-term Tremé and New Orleans residents, we restricted our research to primary and secondary quantitative data.

"Empirical literature on neighborhood stability and these [neighborhood satisfaction] attitudes and behaviors is scant, primarily because it is difficult to collect data sets that contain both individual and neighborhood variables" (Neighborworks, 2011). This study provides information on the individual parcel data along with block and community data that are typically not available in neighborhood change measurement. "While local data could be used, such data would not be comparable or always available across regions" (Cutter et al., 2010, p. 17). "However, the usefulness of quantitative indicators for reducing complexity, measuring progress, mapping, and setting priorities makes them an important tool for decision makers" (Cutter et al., 2008, p. 608). The Tremé study uses integrated data (VGI and municipal sources) at different scales and time periods to provide a more holistic framework from which to measure neighborhood stability. In the case of New Orleans, comparisons of census data from 2000 to 2010 will not reflect the current state of the socioeconomic or demographic reality of the city. Beyond the interruption of Hurricane Katrina, there was a significant change in U.S. Census data collection from the long form to the short form. The more refined or "small data" (Gordon, 2014) previously available from the U.S. Census affects the validity, reliability, and acceptable error of New Orleans longitudinal data. Scholars will benefit from identifying ways to use "middle through" (Ferreira, 1998) data, which are obtained from bottom up (Volunteered Geographic Information [VGI], crowdsourced) and top down (federal census, city blight) sources.

When a disaster occurs, the combination of these data sources makes for a better means to develop short- and long-term planning strategies that will improve the likelihood of neighborhood stabilization. While some may argue that a shift in the demographics from people of color to whites, rise in housing prices, reduction of affordable rental dwellings, reduction in household size, increase in unrelated renters/homeowners, and increase in niche commercial establishments may appear similar to the definition of gentrification, the alternative view can be one of measuring positive growth and measurable effects of neighborhood stabilization. The use of similar variables to define, assess, evaluate, and portray the state of neighborhood change on a continuum should be considered when similar historic neighborhoods, such as Tremé, are being defined using similar matrices but ending up with a planner-defined result.

TREMÉ AND HISTORY OF NEIGHBORHOOD CHANGE

Located directly northwest of New Orleans's French Quarter is the neighborhood known as Tremé, simultaneously one of the most notable and most endangered places in the city. Considered to be the oldest African American neighborhood in the United States, Tremé has been a center of resistance both politically and socially since the earliest days of the French occupation, even before it was established as a neighborhood in 1812 (Campanella, 2008, p. 28). Its rich multicultural history bred many of New Orleans's most cherished traditions, with many scholars tracing the creation of jazz music to the blend of African, Caribbean, and European heritages that mixed unbidden in early Tremé. The neighborhood has a grand tradition of performance, starting with Sunday gatherings in Congo Square. These were the precursors to the second-line parades and colorful Mardi Gras displays, like the Skull and Bones Gang and the gathering of the Mardi Gras Indians, which take place annually. However, decades of disinvestment and damage from Hurricane Katrina have left Tremé vulnerable, and recent trends indicate that much of Tremé's native population is at risk of being displaced, if they haven't been already.

Even before the land was subdivided into the neighborhood we recognize today, it was a place where enslaved Africans and free people of color could meet and exchange goods and ideas. Originally known as the Place de Nègres, the slave market was a gathering place where dance and music was permitted and seasonal events like the Congo Circus were held (Crutcher, 2010, p. 27). While there is little evidence that free people of color outnumbered white Europeans as landowners in the beginning, Tremé would come to distinguish itself as a politically prominent African American neighborhood in the years to come (Campanella, 2008, p. 28). Active in abolition efforts leading up to the Civil War and ultimately instigating the 1896 "separate but equal" Supreme Court mandate in the landmark *Plessy v. Ferguson* decision, residents of Tremé have a long-standing tradition of organization and engagement. However, the twentieth century brought with it numerous slum clearance projects and, consequently, new challenges that residents were not able to repulse (Figure 11.1).

As early as 1926, the city of New Orleans began clearing blocks of Tremé that it deemed to be deteriorating in the name of public works. The first block was demolished to build a new performing arts center, the municipal auditorium, completed in 1929 (Crutcher, 2010, p. 40). The majority of New Orleans's urban renewal took place in the 1960s, as federal funding became available in the form of housing and highway grants. In 1964, over a hundred families were displaced to make way for a new cultural center adjacent to the municipal auditorium. The project was approved and undertaken without any serious long-term planning and was never completed, despite having removed residents from their homes and cleared the land. The majority of the families were relocated to new public housing developments across the river (Crutcher, 2010, p. 44). After many years, the project area was reimagined as Louis Armstrong Park, the gated National Park Service–run heritage park that occupies the southwest corner of Tremé today.

Perhaps the most disruptive of the urban renewal projects was the construction of Interstate 10 through New Orleans between 1961 and 1969. The elevated expressway was built over Claiborne Avenue, a major African American business corridor and recreation area that bisects Tremé and provided much needed public space while many parks and playgrounds remained segregated (Lewis, 2003, p. 98). Locating the expressway over an existing road was favorable when compared with outright slum clearance, but 170 families and 50 businesses were still uprooted to make way for three sets of exit ramps for the interstate (Crutcher, 2010, p. 60). Although its construction is not the only reason accredited to the decline of Claiborne Avenue's livelihood, the following years saw significant disinvestment in the area.

According to the city planning commission, Tremé is the 2 km^2 swath bounded by Rampart Street, Esplanade Avenue, Broad Street, and St. Louis Street. Within those parameters are some of New Orleans's most historic structures, notably St. Augustine's Church, built in 1842. When

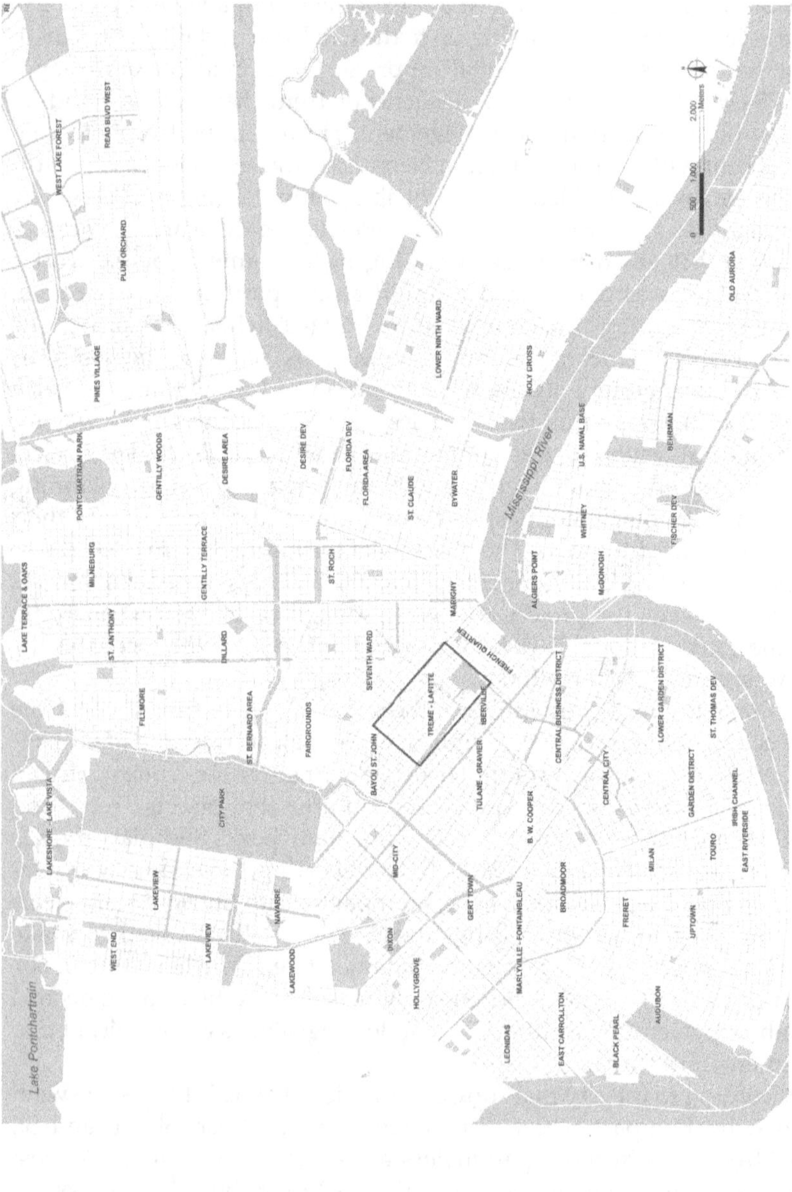

Figure 11.1 Tremé neighborhood within the city of New Orleans, LA. (From Arceneaux, 2014.)

Hurricane Katrina struck New Orleans in August 2005, much of Tremé escaped with comparatively little damage, as it is built on relatively high ground adjacent to the natural levee from the Mississippi. This is not to say that the neighborhood went unscathed, as many homes sustained serious damage from partial inundation and wind, and many still stand empty. As of the 2010 census, Tremé has a population of 4,155, roughly half of its 2000 count. Perhaps more alarming are the demographic shifts we are seeing in conjunction with this population loss—patterns that could potentially change the face of Tremé forever. The Tremé community held strong neighborhood ties and high levels of homeownership, was predominantly African American, has easy access to retail and transportation, and is in close proximity to employment centers. According to Gladstone and Préau, the processes of gentrification, which have already been completed or are nearly completed in all of the surrounding neighborhoods, were in their beginning stages in the early 2000s in Tremé, prior to Hurricane Katrina. Their block-by-block longitudinal study of the areas surrounding the tourist zone, namely, the French Quarter, observed changes in race, owner occupancy, and housing value from 1970 to 2000 (Gladstone and Préau, 2008, p, 146). "During the past three decades, nearly every census block in Tremé lost black residents at a much higher rate than either the study area or the city as a whole, evidence of both housing abandonment in the 1970s and 1980s and displacement of poor renters during the 1990s. Between 1990 and 2000, while the number of black residents declined by 11.4% and the neighborhood continued to lose population, white residents increased by nearly 17%. In parts of Tremé, the increase in white population has exceeded 40%" (Gladstone and Préau, 2008, p. 163). In this study, we hope to update prior findings using a similar methodology to determine what stage in the process Tremé is at currently.

METHODOLOGY AND RESULTS

This chapter aims to conduct an analysis using national and community gathered data sets in order to characterize local circumstances. Community gathered data provide a perspective of on-the-ground conditions within a local context. Since 2010, New Orleans neighborhood groups in conjunction with WhoData.org have been gathering and mapping parcel-level data throughout the city in order to gauge recovery, identify infrastructure problems and blight, and encourage businesses and residents to return. The mapped data are used as a tool by local government regarding recovery progress, unidentified infrastructure problems, and levels of blight. In 2011 and 2013 volunteers from the University of New Orleans, WhoData.org, Project Homecoming, and Providence Housing conducted parcel-by-parcel property condition surveys throughout the Tremé neighborhood utilizing survey methods developed by WhoData.org after Hurricane Katrina. The ability to collect survey data, create condition maps, and communicate the findings could not have been accomplished without developing this GIS implementation plan. Since 2010, the survey methods developed by WhoData.org have been implemented throughout Orleans Parish, collecting information regarding more than 80,000 parcels.

Several meetings were held with WhoData staff and neighborhood volunteers before the project plan was complete in order to ensure that residents were comfortable with the maps and the surveying process itself. The survey teams were then trained to evaluate and collect property condition data using a standardized set of indicators that had been used in neighborhoods across the city. Each of the survey teams received a walking map that was created using Esri's ArcMap, a printed survey sheet, and an overall map of their neighborhood, which denoted where their specific survey sector was. The WhoData team created the walking maps by using the city of New Orleans parcel data. The Tremé was split into survey team sectors, which had roughly the same number of parcels. The data collected by hand on hard-copy survey forms was then imported into a spreadsheet, where it was matched to a specific row based on a unique identifier (GeoPin). When the survey data was finally collected and systematically entered into preformatted spreadsheet databases by residents, it was returned to WhoData for mapping and analysis. This was the one drawback to the current VGI process. There was a significant time commitment that had to be devoted to input the results from 800 parcels with multiple variables. After the survey had been input into a database and imported into ArcMap for analysis, a condition map was created for the neighborhood. The condition map served as a focal point for residents to develop their land use redevelopment plan, focusing on blight, persistent problem areas, neighborhood blocks that are struggling to recover, areas that are emerging, and those with growth potential.

The property metric for buildings is based on a good, fair, or poor status using the WhoData property condition rating system. Good includes buildings with no structural damage, no repairs needed, or that need minor cosmetic work; fair includes buildings with no structural damage, minor repairs needed, or they need significant cosmetic work; and poor refers to a structure with visible structural damage and major repairs needed. Land use refers to whether a parcel is an empty lot or has a permanent structure. Occupancy is based on indicators such as active meters, trash cans, and general signs of an occupied residence.

Statistics regarding building conditions reflect growth and investment within the study area. Between 2011 and 2013, 31% of the surveyed buildings showed improvement, 16% of the buildings showed a digression, and 53% remained the same in condition rating. Unlike the building conditions, the land use patterns and occupancy levels showed minimal change between 2011 and 2013. There were only 3% of land use changes between empty lot and structural presence, while 97% remained the same. Similarly, 16% of parcels showed a change in occupancy; 10% became occupied, while 9% became vacant. These statistics do not show large changes of the landscape and land use within the historic Tremé, but rather, the increase in building conditions implies an increase in investment.

Because the Tremé is categorized as a historic preservation neighborhood and is listed by the New Orleans Historic District Landmarks Commission, demolition, maintenance, and development involve compliance with strict guidelines and are subject to a more stringent permitting process. Initially, "the [urban] pioneers who bought, rehabilitated, and usually reconverted the buildings to single-family occupancy often had difficulty obtaining bank financing for the work necessary to make needed improvements" (Garvin, 2014, p. 332). Many of these pioneers were young urban professionals (YURPs) or suburban transplants who contributed to the "New-Towns-In-Towns" movement. Therefore, many of the improvements were self-financed. This may suggest that building condition could be an indicator of a higher economic class of residents moving into the neighborhood and repairing these historic buildings.

As seen in Figure 11.2, 80% of the parcels within Tremé were identified as having a structure. The structures could be a residential, commercial, church, or mixed-use building. Based upon the VGI survey, 50% of these structures were rated good, 19% fair, and 10% poor. A separate assessment is made on the use of the lots based on if there was a structure or a lot was vacant. Of the total number of lots, 31% of the parcels were identified as vacant.

The building assessment included properties that were rated as occupied or vacant. Properties that are occupied must meet external inspection criteria, such as being secure and having an electric meter in place, a mowed lawn, window treatments, and (if applicable) evidence of transportation. Out of the vacant buildings surveyed, 64% were rated fair, 31% good, and 5% poor. Figure 11.2 summarizes the 2013 results of the

Figure 11.2 Tremé property condition survey results. (From WhoData.com.)

public participation geographic information systems (PPGIS) property condition survey provided by the Historic Faubourg Tremé Neighborhood Association, Project Homecoming, WhoData, and students from the University of New Orleans Department of Planning and Urban Studies.

Data from the 2000 and 2010 U.S. Census were also compared and analyzed at the block group level to supplement the findings of the crowd-sourced parcel survey. Census data at the household level remain restricted, and therefore block-level data are the most relevant and readily available to the public. Tremé contains four census tracts split into nine block groups, although the entirety of Census Tract 44.02, which encompassed the Lafitte Housing Project, was cleared for demolition in 2008. The Lafitte remains unoccupied and skews the data for Tremé as a whole; therefore, Census Tract 44.02 will not be included in most of the statistical analysis. Regardless, it is important to consider the implications of the demolition of this project, which was 99% African American occupied in 2000, to make way for mixed-income housing in a centrally located and in-demand area.

In 2010, Tremé's population was 4,155 total residents—less than half of its 2000 total of 8,869. The demolition of Lafitte alone accounts for the loss of 2,622 residents from the area, with the remainder divided among the seven other census tracts in the neighborhood. Excluding Lafitte, the block groups averaged a loss of 299 people each, although the block groups bordering the French Quarter in Census Tract 39 lost far fewer than the other block groups. However, it is the demographic and economic changes that occurred since 2000 that have given a better indication of the direction of neighborhood change in the Tremé. In order to determine whether the native population of Tremé is being displaced, several population characteristics and changes were observed from 2000 to 2010. These characteristics included race, household size, age, and renter versus owner occupancy.

For the past century, Tremé was primarily an African American neighborhood. As of 2010, it was majorly African American at 75.3%, although this is considerably less than the 93.6% it was at in 2000, or 91.3% excluding Lafitte. Further, Tremé has experienced an uneven shift in racial change. The percentage of African American residents in the two block groups within Census Tract 39 declined by an average of 29.3%, while the block groups in Tracts 40 and 44.01 (which are on the other side of the Claiborne Expressway) only declined 7.77%. Conversely, the percentage of white residents increased by 24.8% in the block groups of Census Tract 39, but only an average of 5.6% in the block groups in Tracts 40 and 44.01. The Hispanic population has also increased since 2000, by 8.2% in Tract 39 and 2.8% in Tracts 40 and 44.02 (Figure 11.3).

To capture the shift from large families occupying rented homes to smaller households of young professionals with buying power, 2000 and 2010 levels of renter occupancy, household size, and age were compared. The percentage of renter-occupied homes decreased an average of 5.9% in the neighborhood, although in this instance there was not a great disparity between the census tracts on the two different sides of Claiborne Avenue; Tract 39 decreased 6.8% in renter occupancy and Tracts 40 and 44.01 decreased an average of 5.5%. Median age increased in the 10-year period by an average of 6.2 years in the three tracts. This characteristic also didn't demonstrate a large difference between Tracts 39, 40, and 44.01, averaging 5.6-, 6.1-, and 6.6-year increases in median age, respectively. The percentage of households with more than three occupants was used to determine whether the larger, multigenerational families had been driven out in favor of smaller ones. The Lafitte Housing Project had primarily large households in 2000, with 54.3% consisting of three or more inhabitants. Excluding Lafitte, the decrease in households hosting over three people is 7.1% on average, also without a significant difference in the three tracts.

The results of this comparative study suggest that there is a positive correlation of factors that suggest both gentrification and neighborhood

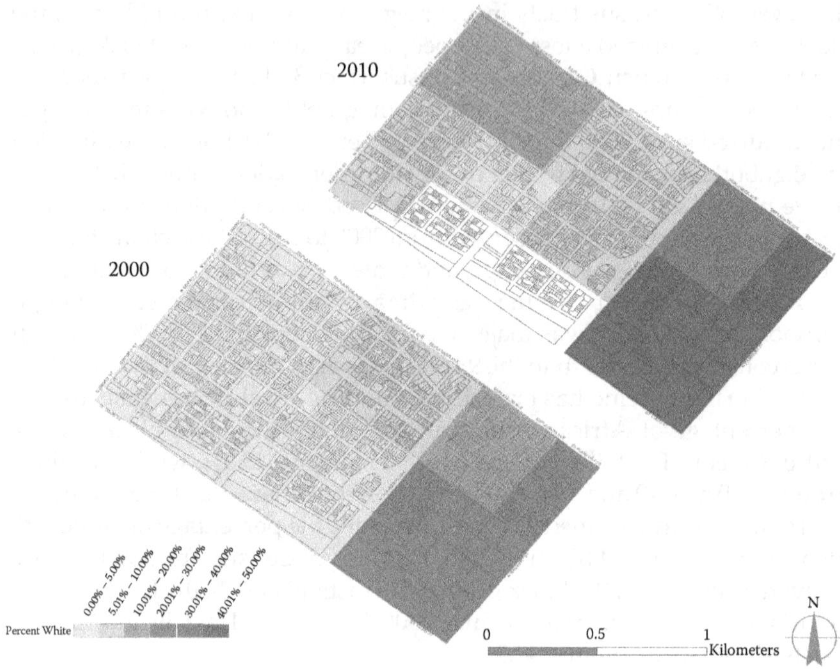

Figure 11.3 Change in percentage of white residents in Tremé, 2000–2010. (From Major, 2014.)

stabilization at the same time. Specifically, there has been a shift in the race of neighborhood citizens from primarily African American to white; the building condition improved overall, with a larger percentage surveyed as good; and the price of housing and rents have increased, while lot vacancy rates and family size both decreased in the 10-year period. After examining the combination of data at varying scales, and using the definition provided earlier, the results of this study suggest that Tremé has moved from a neighborhood recovering from a disaster to a more stable neighborhood.

LIMITING CONDITIONS

There was no data aggregation or weighting, as this is another model for integrating parcel-level data that has collection standards to measure and triangulate perception, municipal/public policy, and community knowledge. While methods exist for determining weights that are subjective or data reliant, such weighting schemes do not always reflect the priorities of decision makers (Esty et al., 2005; Cutter et al., 2008).

Due to a fluid landscape within the historic Tremé neighborhood, VGI or crowdsourced data were used to summarize statistics only for buildings that existed in both 2011 and 2013. In 2011, buildings made up 48% of parcel land use, while in 2013, buildings made up 80%. The statistics included in the collection of VGI data reflect patterns within the neighborhood, but do not account for all parcels within the historic Tremé. Forty-three percent of the 2011 data are missing due to property record changes, modified city parcel layers, and incomplete surveys. The statistics in the section only account for those 1,254 parcels that were surveyed in 2011 and 2013.

CONCLUSION

Planning literature classifies neighborhoods in flux and measures their resilience, state of gentrification, or level of neighborhood stabilization using similar models and metrics. Many of the same variables are used to measure community status. This study focuses on defining a contemporary measure of neighborhood stabilization after a man-made or natural disaster using crowdsourced and municipal (integrated) data. The DROP model was applied to the case study of Tremé and included municipal neighborhood boundary definitions, standardized property condition assessments, and small sociodemographic, integrated temporal, and scalable data. It is important to involve residents in the type of community data collected and agree upon the definition, standards, dissemination, and goals for use prior to using a VGI model. The model results can aid residents in making informed planning decisions, but not when a neighborhood and its residents remain in shock after traumatic events. This research can be used to establish a baseline for neighborhood change and a model that can be applied for older cities, whether or not they are recovering from a catastrophe such as Hurricane Katrina in 2005.

Traditional planning education requires a more contemporary approach to evaluating the state of a neighborhood after a man-made or natural disaster. Many of the data sources used in longitudinal studies rely on U.S. Census data, which has, in recent years, decreased the availability of small data. City databases that may include demographic profiles with less margin of error are typically not accessible or in compatible formats. Variables, such as race, age, household size, property value, and housing condition, are used, while the level of geography varies. This research considers ways to create an advanced model for neighborhood planning analysis. The focus of this research portrays neighborhood stability using national data sets, specifically the U.S. Census, while integrating VGI, which is typically not available.

A review of literature outlines similar measures to monitor gentrification and determine neighborhood stability. The definitions focus on demographics from African American to white, lower income to higher, improved property conditions, a reduction in blight and vacancies, and housing multigenerational families or single or unrelated couples. The conceptual framework established in this chapter focuses on a neighborhood level of analysis. The research framework used analytical techniques that are easily replicable, data that are freely available, and spatial analysis that is based on industry standards.

Using geographic information systems, comparative statistics of demographics, housing condition, and ownership were provided. Based upon the findings, progress toward neighborhood stability was exemplified in the shift in specific variables such as race, where the population shifted from primarily African American to white; income, which increased significantly from lower to higher; and homeownership and age, which increased. The status of property and use condition received an added benefit from using VGI collected through collaboration with a Tremé resident association, student volunteers, and university and neighborhood organizations. While other studies include weighted overlays of data, this model provides a way to use readily available public data that individually or collectively demonstrate trends and factors that contribute to the stabilization of neighborhoods and should be used by policy makers in a postrecovery context.

By using methodology that uses U.S. Census data, the ability to identify and compare neighborhood change is replicable. This data-driven model integrates public and crowdsourced or VGI data. In this case, the VGI used the WhoData model for data collection, which can integrate with other New Orleans data sets. The results of this, and the former Gladstone–Prieu study, suggest that this neighborhood has been in the process of gentrification since the 1970s. While there are negative connotations related to gentrification, the variables used to measure this phenomenon are similar to those for defining neighborhood stabilization. Scholars and planners in practice should consider the impact of applying labels of neighborhood cycles that are evaluated using the same metrics. These labels can affect the resident perception and the ability of a neighborhood to emerge, grow, revitalize, or stabilize through public and private investment.

Future research should consider how the data used to evaluate different phases of the neighborhood life cycle can be used in an integrated study using spatial data at varying levels of geography. In summary, the research presented for Tremé suggests a conceptual framework for neighborhood planning that is not complex, is replicable, and allows for empirical analysis of communities that aids in policy, practice, and theoretical study of neighborhood stability.

REFERENCES

AllGov. (2014). All things our government really does. NeighborWorks America. http://www.allgov.com/departments/independent-agencies/neighborworks-america?agencyid = 7447 (accessed December 5, 2014).

Arceneaux, B. (2014). Tremé neighborhood boundary. Digital image [web].

Campanella, R. (2008). *Bienville's Dilemma: A Historical Geography of New Orleans*. Lafayette: Center for Louisiana Studies, University of Louisiana at Lafayette.

Crutcher, M. (2010). *Treme Race and Place in a New Orleans Neighborhood*. Athens: University of Georgia Press.

Cutter, S. L., Barnes, L., Berry, M., Burton, C., Evans, E., Tate, E., and Webb, J. (2008). A place-based model for understanding community resilience to natural disasters. *Global Environmental Change*, 18, 598–606.

Cutter, S. L., Burton, C. G., and Emrich, C. T. (2010). Disaster resilience indicators for benchmarking baseline conditions. *Journal of Homeland Security and Emergency Management*, 7(1), article 51.

Federal Geographic Data Center. www.fgdc.gov.

Ferreira, J. (1998). Information technologies that change relationships between low-income communities and the public and non-profit agencies that serve them. In *High Technology and Low Income Communities: Prospects for the Positive Use of Advanced IT*. Cambridge: MIT Press.

Gladstone, D., and Préau, J. (2008). Gentrification in tourist cities: Evidence from New Orleans before and after Hurricane Katrina. *Housing Policy Debate*, 19(1), 137–175.

Global Facility for Disaster Reduction and Recovery. (2014). *Crowdsourced Geographic Information Use in Government*. Washington, DC: World Bank.

Gordon, S. (2014). 5 ways small data can be more valuable than big data: Real-time information lets your business act more quickly. August 22. http://www.destinationcrm.com/Articles/Web-Exclusives/Viewpoints/5-Ways-Small-Data-Can-Be-More-Valuable-than-Big-Data-98927.aspx (accessed December 6, 2014).

Lauria, M., and Stout, M. E. (1995). The significance of scale in the analysis of gentrification. Working papers, paper 9. New Orleans, LA: College of Urban and Public Affairs (CUPA), University of New Orleans.

Lewis, P. (2003). *New Orleans: The Making of an Urban Landscape*. 2nd ed. Sante Fe, NM: Center for American Places in association with the University of Virginia Press.

Major, E. (2014). Non-Hispanic White population change from 2000 to 2010. Digital image [web].

Mallach, A. (2008). *Managing Neighborhood Change: A Framework for Sustainable and Equitable Revitalization*. Montclair, NJ: National Housing Institute.

Metzger, J. T. (2000). Planned abandonment: The neighborhood life-cycle theory and national urban policy. *Housing Policy Debate*, 11(1), 7–40.

Neighborworks. (2011). *Evaluating Community Stabilization Efforts at the Neighborhood Level: A NeighborWorks® Pilot with Success Measures*. Washington, DC: Neighborhood Reinvestment Corporation.

Pezzullo, P. C. (2010). Tourists and/as disasters: Rebuilding, remembering, and responsibility in New Orleans. *Tourist Studies*, 9(1), 23–41.

Rawlings, L., Harris, L., and Turner, M. A., with Padilla, S. (2011). *Race and Residence: Prospects for Stable Neighborhood Integration*. Washington, DC: Urban Institute.

Rohe, W. M., and Stewart, L. S. (1996). Homeownership and neighborhood stability. *Housing Policy Debate*, 7(1), 37–80.

Roussey, S. (2013). Enhancing resilience to shocks and stresses. Briefing paper. ACF-International, Paris, France, April.

Rybarczyk, G., and Mohapatra, R. P. (2013). Evaluating neighborhood through empirical analysis and geographic information systems. *Journal of the Urban and Regional Information Systems Association*, 25(1), 63–75.

Stewart, W. R. (1996). Homeownership and neighborhood stability. *Housing Policy Debate*, 7, 46.

Tufts University. (2009). *For Dudley, by Dudley: An Analysis of Gentrification Risk in the Dudley*. Boston: Tufts University, Urban and Environmental Policy and Planning.

WhoData. www.whodata.net.

12

Clustered and Community-Driven Housing Recovery

Lessons Learned from Hurricane Katrina to the Great East Japan Earthquake

Tamiyo Kondo

Contents

INTRODUCTION

The catastrophic aftermath of Hurricane Katrina presented an enormous challenge of rebuilding along the Gulf Coast. In August 2005, Hurricane Katrina's storm surge washed over the levees and breached the floodwalls of New Orleans. Water stood in more than 80% of the city for nearly 2 weeks, and 180,000 housing units were severely damaged or destroyed [1] (Table 12.1). Katrina resulted in the largest U.S. displacement of residents throughout the country. There is one other disaster very similar to this situation: the 9.0 magnitude Great East Japan Earthquake of 2011, which set off a devastating tsunami that sent walls of water washing over coastal cities in the northern part of Japan.

Hurricanes and earthquake/tsunami hazards are different, but there are many similarities between these disasters in terms of social phenomena, such as the high percentage of submerged area, widespread and long-term evacuation of survivors, and vast extent of destruction of the built environment (Table 12.1). If the extent of damage and the social trends are similar, the challenges ahead for long-term recovery will have something in common. This implies that Japan could learn from the experience of Hurricane Katrina and its failures in order to achieve a more sustainable disaster recovery for ongoing long-term postdisaster recovery processes and planning.

This chapter discusses housing recovery after two separate natural disasters: New Orleans after Hurricane Katrina and the Great East Japan Earthquake. The author defines housing recovery not only as the act of house reconstruction, but also as a decision-making and reconstruction process that restores people's living. In this chapter the author attempts to convince readers that the concepts of clustering and community-driven

Table 12.1 Hurricane Katrina and the Great East Japan Earthquake

	Hurricane Katrina	**Great East Japan Earthquake**
Date/year	August 2005	March 2011
Hazard	Hurricane, wind, rain	Earthquake, tsunami, nuclear accident
Disaster-stricken area	• Gulf Coast (Louisiana and Mississippi) • Greater New Orleans (New Orleans Parish, St. Bernard Parish)	• Tohoku Region Coast along the Japan Sea • Prefectures of Aomori, Iwate, Miyagi, Fukushima, Ibaragi, and Tokyo
Human damage	More than 1,800 deaths	18,958 deaths, 2,655 missing

Housing damage	• Alabama, Florida, Louisiana, Mississippi, Texas: 1,197,400 buildings
	• New Orleans: 80% of all the buildings sustained some damage from high winds and water [2]
	• New Orleans: More than 100,000 homes (half of the city's total) were under more than 4 feet of floodwater [2]
		• Miyagi Prefecture: 460,826 buildings
		  • Higashi-Matsushima: 79.8% damaged
		  • Yamamoto: 68.4% damaged
		  • Minami-Sanriku: 64.6% damaged
		• Iwate Prefecture: 44,307 buildings
		  • Otsuchi: 68.2% damaged
		  • Rikuzentakata: 43.2% damaged

Flooded area	• 374 km^2 flooded in New Orleans • 80% flooded in New Orleans	• 560 km^2 flooded in northern Japan • 65% flooded in Higashi-Matsushima • 52% flooded in Otsuchi

housing are especially important for recovery after large natural disasters. The author defines clustered housing recovery as the housing reconstruction process controlled and guided by the spatial planning method, including infill development and land use planning. The importance of the clustered housing recovery model was proven by what happened after two catastrophes, such as the checkerboard housing recovery at the neighborhood level after Hurricane Katrina and individual housing reconstruction with relocation at the city level after the Great East Japan Earthquake, which will be described in this chapter. The community-driven approach is one that implements clustered housing recovery.

This chapter is based on continuing field surveys of housing rebuilding and interviews with homeowners in New Orleans after Hurricane Katrina, and Miyagi and Iwate Prefectures after the Great East Japan Earthquake of 2011.

HOUSING RECOVERY AFTER HURRICANE KATRINA IN NEW ORLEANS

After Hurricane Katrina, what kind of housing recovery has been implemented in New Orleans since 2005? Did people come back to their neighborhood where they used to live before Katrina hit? What was the challenge of housing recovery after such a catastrophe?

Housing Damage and Population Recovery by Neighborhoods

Hurricane Katrina was one of the strongest storms to impact the coast of the United States in the last 100 years. The loss of life and property damage were caused not only by strong winds and rainfall, but also by breaches of the levees that separate New Orleans from surrounding bodies of water. The combination of strong winds, heavy rainfall, and storm surge led to breaks in the earthen levees after the storm passed, leaving some parts of New Orleans under 20 feet of water [2].

Figure 12.1 shows the housing damage by neighborhood in New Orleans. Almost 80% of the city was flooded (drawn as dots) and experienced some sort of damage, but the extent of housing damage by neighborhood is quite different; 21 out of 73 neighborhoods in the city experienced more than 40% damage of the total housing stock in each neighborhood. This housing damage is calculated as the average housing damage (%) of housing that was assessed by the Federal Emergency Management Agency (FEMA). The estimates of housing unit damage are based on direct inspection of housing units by FEMA to determine eligibility for FEMA housing assistance. FEMA inspects properties to assess eligibility for real property and personal property assistance; real property assistance is determined as the cost to make repairs to make the home habitable [1].

Figure 12.2 shows the percentage of post households receiving mail by neighborhood in 2014 [3]. Trends in residential addresses receiving mail serve as a useful indicator of the rate of change in occupied housing units; however, active residential addresses are not the same as occupied

Figure 12.1 Housing damage by neighborhood in New Orleans.

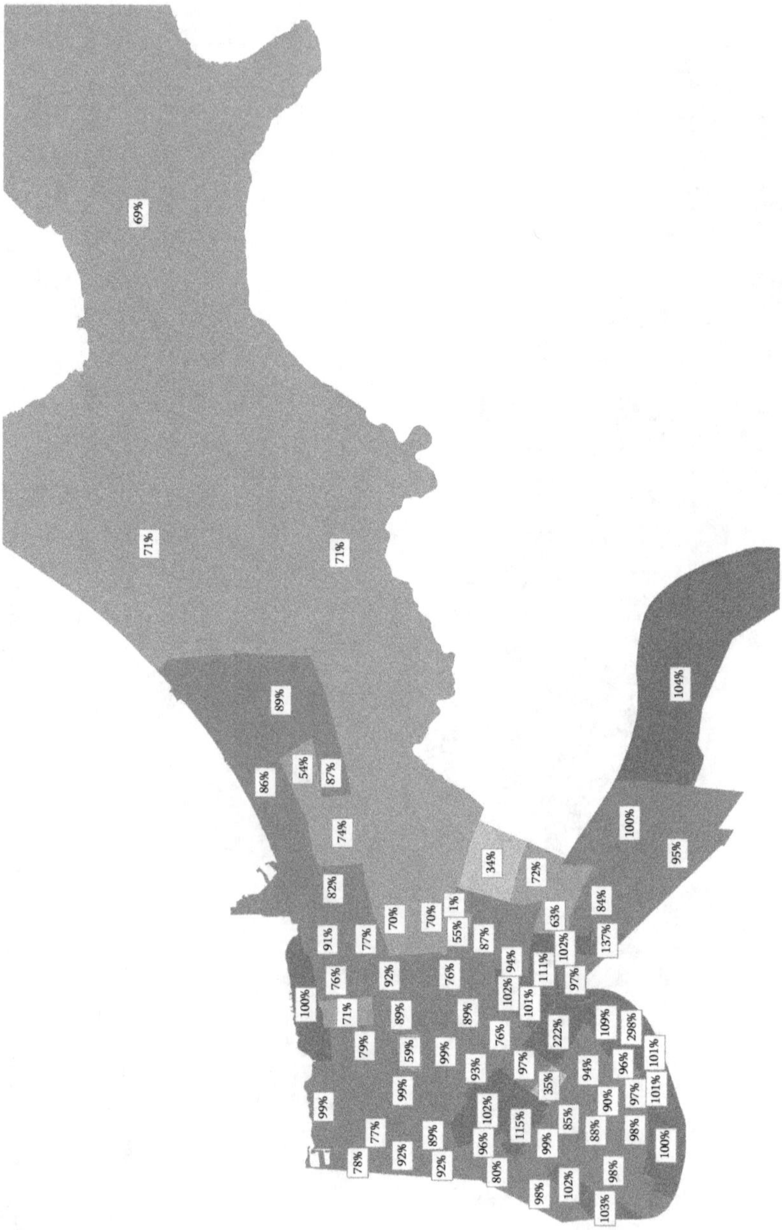

Figure 12.2 Households receiving mail by neighborhood in 2014.

housing units [3]. This is because the U.S. Postal Service (USPS) defines a residential address as actively receiving mail if mail has been picked up within the previous 90 days [3]. We cannot understand actual change and progress of the housing reconstruction situation and its spatial distribution at the neighborhood level according to these data. This was a starting point of the author's housing rebuilding survey in 2009.

Land Use Control by Local Government

Land use control policy is a significant measure to decrease disaster risk, such as not allowing buildings to be constructed near active faults for potential earthquakes or in land below sea level for potential flooding. At the same time, it considerably impacts people's decision-making process regarding where to live and reconstruct their housing after disaster. What kind of land use control policy did New Orleans prepare after Katrina?

The first land use control policy made public by local government was known as the Green Dot Map, which appeared in a local newspaper, the *Times Picayune*. This map is based on the postdisaster recovery plan developed by the Bring New Orleans Back Commission, headed by the mayor of New Orleans at the time. Green dots are shown in areas expected to become parks and green spaces, which made people think these areas would be bulldozed, and therefore they could not come back to their home [4]. In response to the outrage and panic of residents, local government leaders did not discourage people from moving back to vulnerable areas in the city.

The only countermeasures to land use control after Katrina in New Orleans are the base flood elevation regulations determined by FEMA, which specified building elevation levels required for new housing construction. Most people assumed there would be significant changes in the elevation rules following Katrina; however, FEMA announced in April 2006 that base flood elevations would remain unchanged [5]. It is assumed that the federal government concluded that the failure brought about by Katrina was not related to the land use control, but caused instead by the collapse of the levees. In New Orleans, nonstructural disaster reduction measures were not improved after experiencing Katrina [6]. The decision of where to live was left to people at the individual level.

Road Home Program by State Government

What kind of housing subsidy program was prepared for homeowners to rebuild their housing after Katrina? Because of the levee failure, the state of

Louisiana decided to provide housing rebuilding compensation for homeowners through the Road Home Program. Eligible homeowners could receive up to $150,000 in compensation for their losses to help them get back into their homes. There were many problems related to this program, such as the delays in the payment of subsidies, the decrease in urban population resulting from relocation to areas elsewhere in the state or out of state due to the options offered under the program, and the disadvantages under the program for low-income people because the compensation grants were determined by both housing damage and prestorm property value. Green and Olshansky [7] concluded that "the Road Home Program was successful in funneling billions of dollars to homeowners for rebuilding and mitigation, but that the unresolved tension between its role as a rebuilding program and a compensation program created significant barriers to recovery."

Figure 12.3 shows the spatial distribution of property sold to the government by the Road Home Program. The program provided compensation options for homeowners based on three choices: (1) *stay,* for homeowners that wanted to stay in their homes; (2) *relocate,* for homeowners that wanted to sell their homes but remain homeowners in Louisiana; and (3) *sell,* for homeowners that wanted to sell their homes and either move out of Louisiana or

Figure 12.3 Spatial distribution of property sold to the government by the Road Home Program.

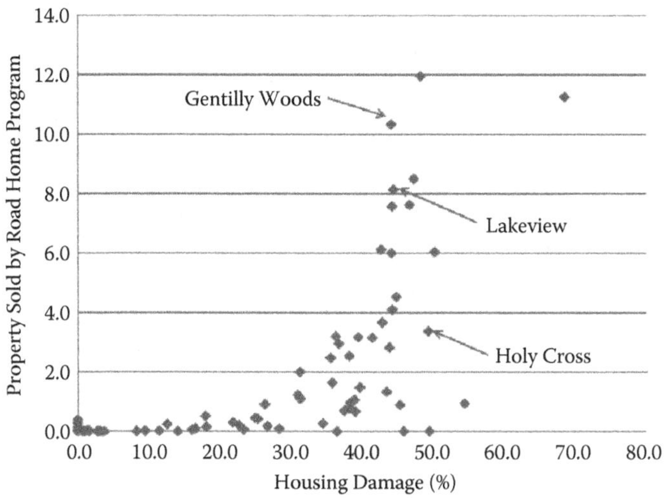

Graph 12.1 Housing damage and property sold by neighborhood.

remain in the state, but as a renter instead. The dots in Figure 12.3 show the distribution of homeowners who chose the second and third options. Out of all the program applicants, the share of homeowners who chose these options was 87.8% in the state and 84.6% in New Orleans [8].

Graph 12.1 shows both the housing damage and property sold by neighborhood. It indicates that there is a positive correlation between properties sold to the Road Home Program and housing damage; however, despite similar levels of housing damage, there is a big difference between the three neighborhoods where the author conducted the housing reconstruction survey shown in the following section. Which elements influenced homeowners' decisions to stay or leave? How did this government subsidy program help survivors come back to New Orleans and rebuild their housing?

Housing Reconstruction Survey

The author started an annual housing rebuilding field survey in September 2009, on the fourth anniversary of Hurricane Katrina, and covered more than 1,500 housing units in New Orleans. The objective of the survey was to clarify neighborhoods' gaps in terms of housing rebuilding speed and property sales [8]. "Why is the speed and situation of housing recovery different?" is an example of a question on the survey. Three neighborhoods were selected: Lakeview, Gentilly Woods, and Holy Cross. All three had similar levels of housing damage. It is not surprising that if damage is massive, the speed of housing reconstruction is slow. The author wanted to know what kinds of factors widen the gap in housing rebuilding and property sales by neighborhood. This is why the author selected neighborhoods that had different characteristics in terms of income, race, and home ownership. Figure 12.4 shows the location of three neighborhoods that are close to the levee breach. Approximately 20% of all properties in each neighborhood were selected as the target of the survey.

The survey was conducted in 2009, 2010, 2012, and 2013 by the author and students from Kobe University who had visited New Orleans annually every September. Figure 12.5 depicts the methodology for carrying out the housing reconstruction survey. The investigator, with a camera and reporting sheet, took a photo of each property, recorded the situation of housing reconstruction, and classified it into one of four categories: reconstructed, under construction, abandoned, or vacant lot. *Abandoned* housing is unoccupied and unmanaged property, which is the most delayed housing reconstruction situation among the four categories. In

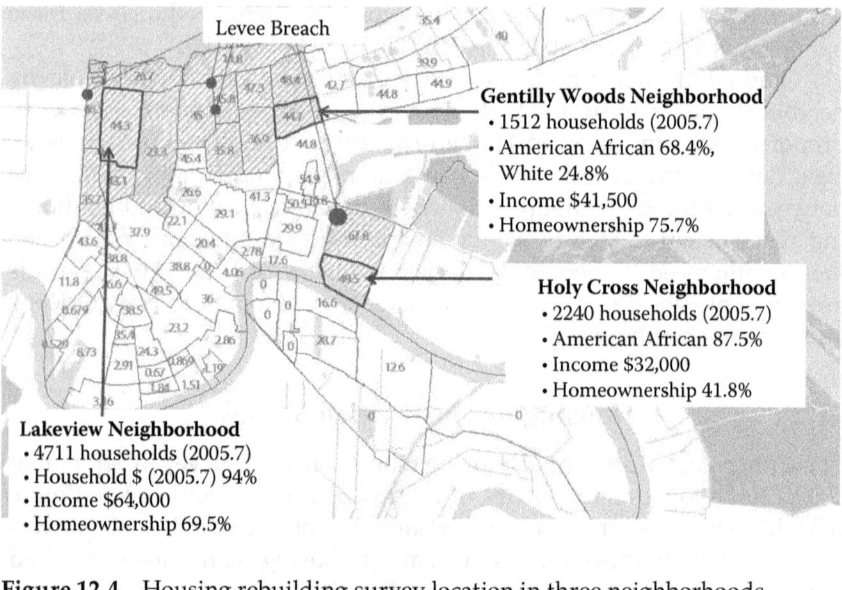

Figure 12.4 Housing rebuilding survey location in three neighborhoods.

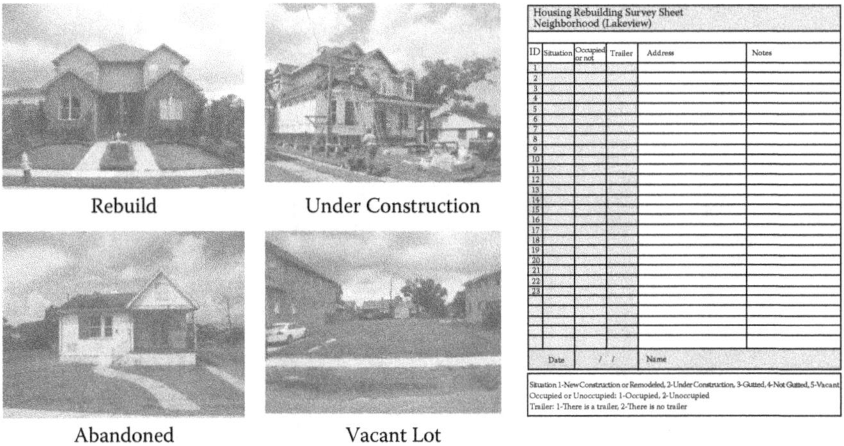

Figure 12.5 Methodology of housing rebuilding survey.

general, the abandoned property is found in such a state that the property owners have intentionally given up control of the property, with no intent of returning to or recovering the property. In comparison with vacant, the homeowners of abandoned properties have not taken any perceivable actions related to a decision on whether or not to reconstruct housing. *Vacant* is considered to be the consequence of a decision to reconstruct housing again or sell the property to others.

When looking at Figure 12.2, the population recovery rate in three neighborhoods does not show such a large difference. Does this mean that the progress of housing reconstruction results in somewhat similar situations?

Gap of Housing Reconstruction in Three Neighborhoods

Graph 12.2 shows the housing reconstruction gap by percentage basis between neighborhoods in 2009 and 2013. The most distinctive trend in 2009 was that the housing reconstruction percentage in all neighborhoods hovered around 50%, and there was a significant difference in the percentage of abandoned housing and vacant lots [8]. What is different is that the share of abandoned properties is approximately 30% of all properties in the Holy Cross and Gentilly Woods neighborhoods; however, approximately 30%

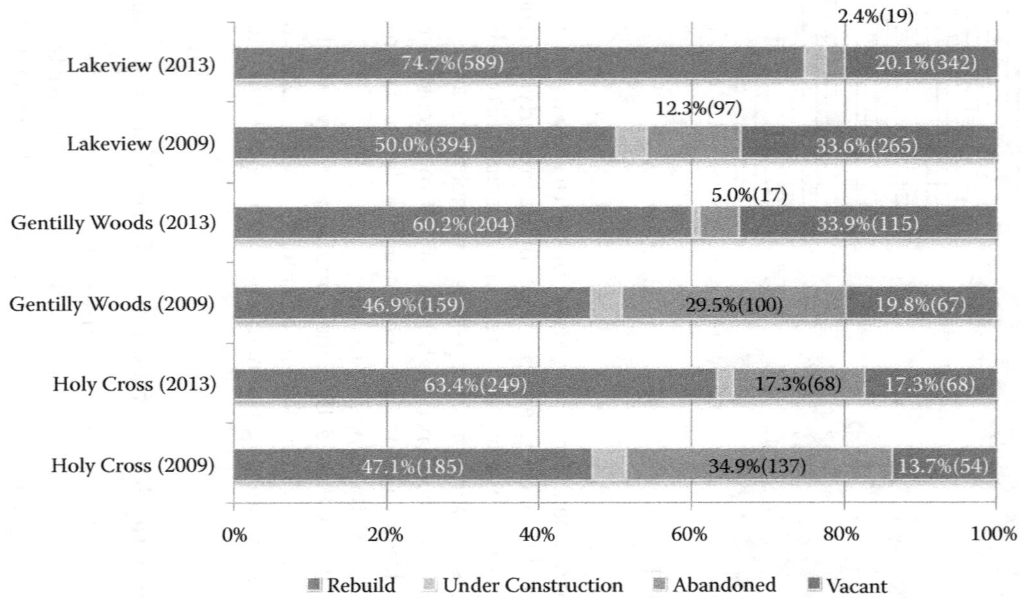

Graph 12.2 Housing reconstruction gap between neighborhoods in 2009 and 2013.

are already vacant in Lakeview. It is clear that by the fourth anniversary of Hurricane Katrina, a significant gap by neighborhood had already emerged.

In 2013, more than 75% of properties were already reconstructed or under construction in Lakeview. Compared with Gentilly Woods and Holy Cross, the 4.3% percentage gap of reconstruction and under construction in Lakeview is small; it is 65.4% and 61.1% in the other neighborhoods. However, at 34.9%, Holy Cross had the largest percentage of abandoned lots, in contrast to Gentilly Woods (29.5%) and Lakeview (12.3%). The neighborhood without a decreasing number of abandoned properties and with an increasing number of vacant lots was found in Holy Cross, and this indicated future widening disparity between neighborhoods.

Figures 12.6 and 12.7 show the housing reconstruction situation and spatial distribution in three neighborhoods between 2009 and 2013. Vacant (white) and abandoned (dark gray) lots decrease year by year and turned into reconstructed or under-construction properties. When looking at the

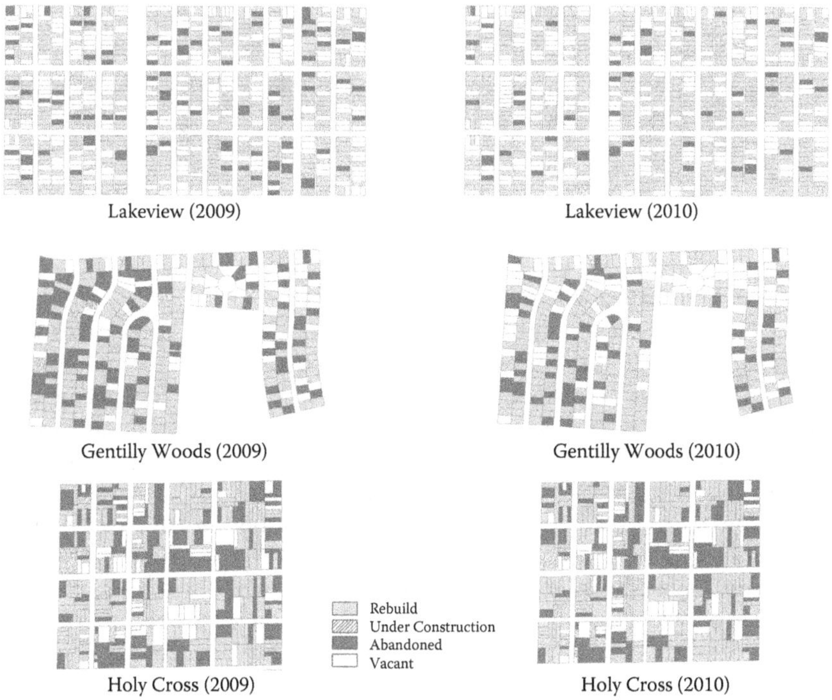

Figure 12.6 Housing reconstruction and spatial distribution in 2009 and 2010.

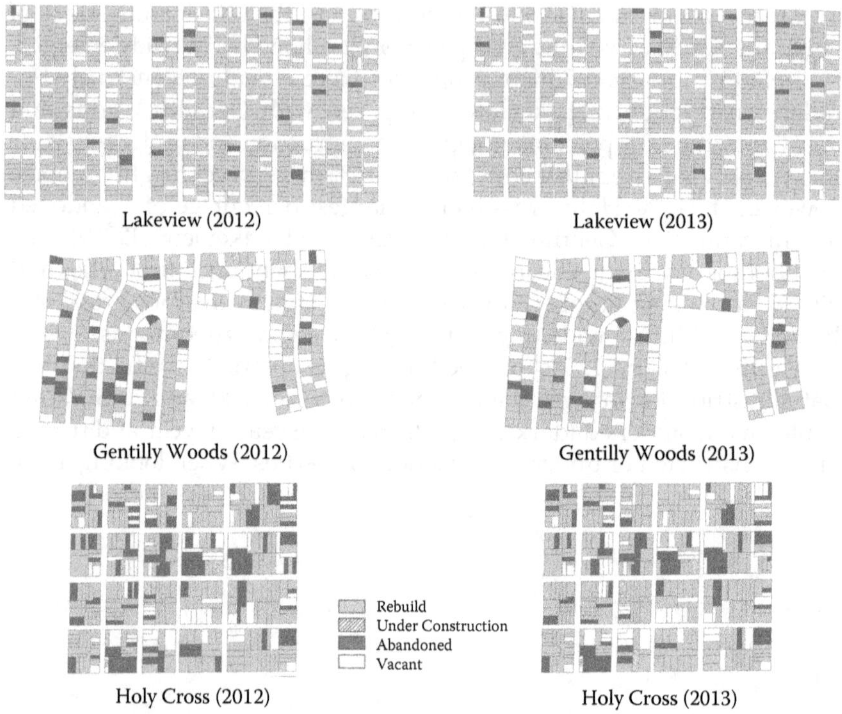

Figure 12.7 Housing reconstruction and spatial distribution in 2012 and 2013.

annual change in each neighborhood shown visually in Figures 12.6 and 12.7, it is obvious that the progress of housing reconstruction is fastest in the Lakeview neighborhood. The housing rebuilding and reconstruction was not finished within 4 years of the disaster; in fact, progress is still ongoing as of the 8-year anniversary. The progress between 2009 and 2013 is about the same in Lakeview and Gentilly Woods, but the progress in Holy Cross is the slowest. Is this because the situation in 2009 already contained a gap in housing reconstruction?

Figure 12.8 shows the positive change properties, which include under reconstruction to reconstructed, abandoned to reconstructed, under reconstruction and vacant, and vacant to reconstructed and under reconstruction. The percentages of positive change properties among all properties in each neighborhood are 30.8% in Lakeview, 30.4% in Gentilly Woods, and 22.6% in Holy Cross. These indicate that the housing

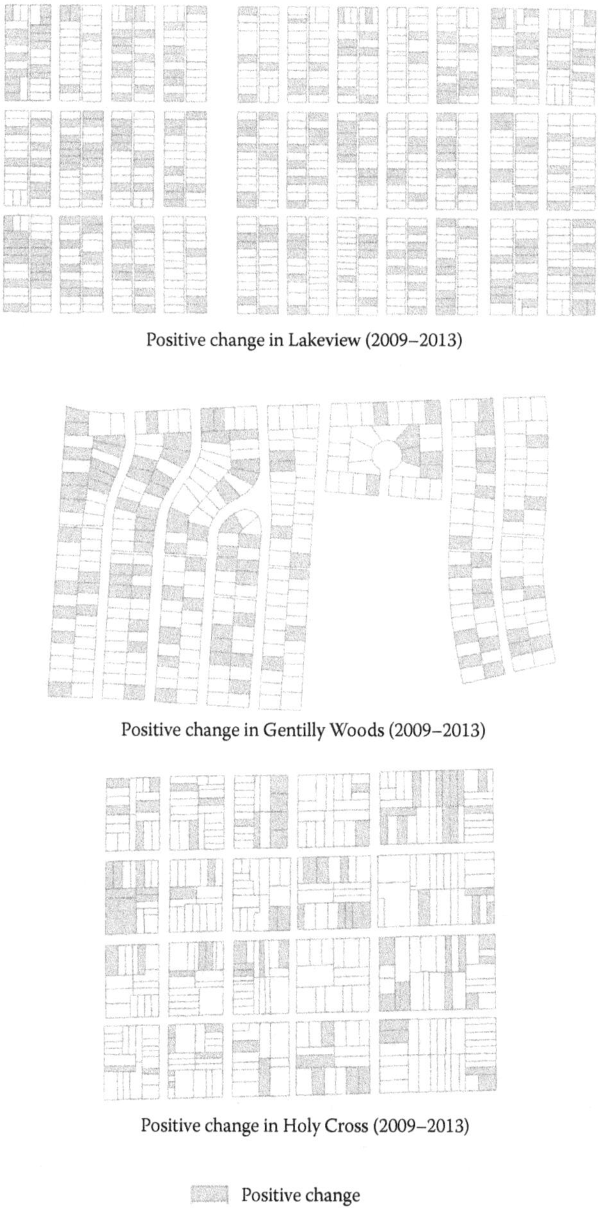

Figure 12.8 "Positive change" properties in three neighborhoods.

reconstruction progress is affected not only by the situation at the starting point, but also by the speed of annual change.

Gentrified or Thinly Populated

Figure 12.9 shows properties that merged after Hurricane Katrina. This is considered a consequence brought about by the local government's Lot Next Door program. The ownership of properties that were acquired as part of the Road Home Program was transferred to the New Orleans Redevelopment Agency (NORA), which also received ownership of blighted properties seized before and after Katrina. Blighted is different from abandoned in the sense that it is a legal term for property that is in an abandoned condition. The Lot Next Door program provided an opportunity for property owners who shared a common boundary with a property owned by NORA to purchase this neighboring property. It is obvious that the Lakeview neighborhood was transformed into a lower-density residential area with large yards and pools. The Lot Next Door program is not designed to return residents to the city, but to ensure that vacant lots or abandoned properties are not left unmanaged. The social recovery, defined as the status of former residents returning home, is unknown at this time. The author found many "for sale" and "for lease" properties in Lakeview, which implies residential turnover. The 2010 U.S. Census shows that the number of elderly people is decreasing and the younger generation is increasing in Lakeview.

The story of housing recovery in each neighborhood can be assumed as follows. In the Lakeview neighborhood, the housing reconstruction is a step ahead of that in other neighborhoods. In addition to the high percentage of reconstructed property, there are many positive vacant lots that are expected to be reconstructed or sold. It is expected that senior residents will sell their properties, and then younger families will begin moving into Lakeview, an elite neighborhood, where the property values have slightly declined because of the flooding. Relatively high-income populations can participate in the government Lot Next Door program, which has gentrified the neighborhood. In Gentilly Woods, the Road Home Program functions to decrease abandoned property.

In Holy Cross neighborhood, there are many abandoned properties, which means that there are many residents who have not made a decision on whether to come back to the neighborhood. The number of applicants of the Road Home Program is relatively small in Holy Cross, and it is expected that much of the housing cannot be reconstructed after disaster.

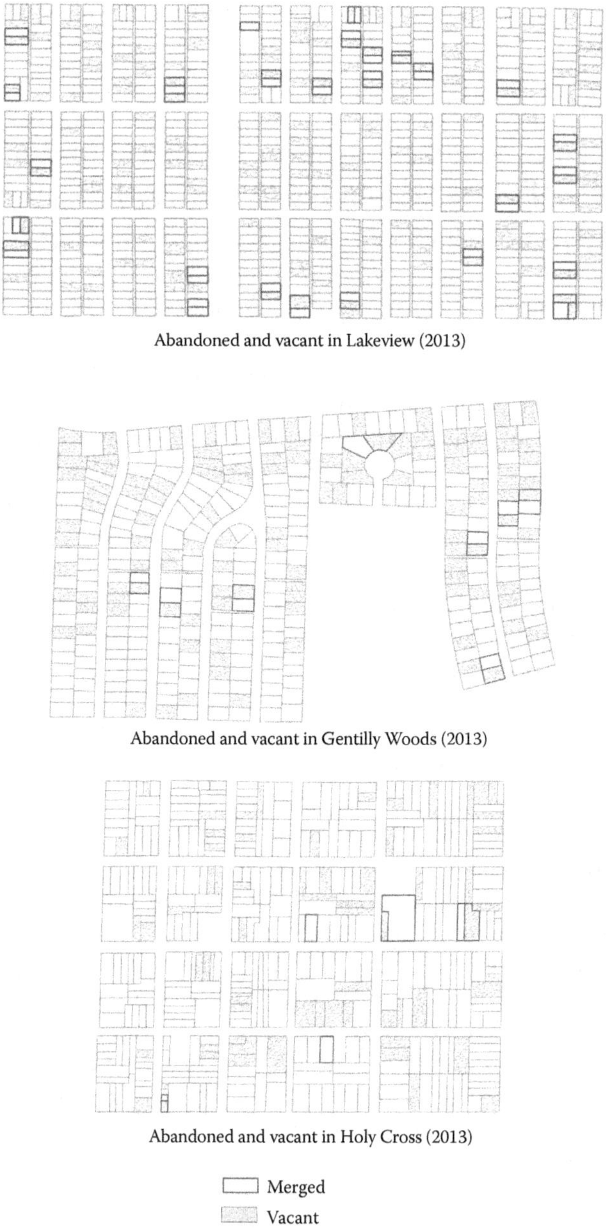

Figure 12.9 Merged properties after Hurricane Katrina.

There remain a number of abandoned properties that will cause deterioration of the dwelling environment and difficulties for residents returning to their neighborhoods [9].

What Caused the Neighborhood Gap in Housing Reconstruction and Property Sales?

What caused the neighborhood gap in housing reconstruction, property sales, and turnover of the neighborhood population? In general, it is well known that vulnerability increases the likelihood of difficulties in recovering from the negative impacts of hazardous events. The vulnerability of a neighborhood is composed of individuals' income and home ownership, property values, and the community.

As expected, income and homeownership have a significant impact on housing reconstruction. The Road Home Program works by reimbursing homeowners for housing damage experienced during Hurrican Katrina. Reimbursement for low-income homeowners is small in comparison to middle- and high-income residents. This is because compensation was calculated by the prior property values as well as factoring in the damage from the storms. In addition to income, home ownership is assumed to have an impact on housing reconstruction. This is because tenants cannot participate in housing reconstruction, and landlords don't have great incentive to reconstruct their rental properties. The accumulation of individuals' income and homeownership affect housing reconstruction in neighborhoods.

The other vulnerability that might have influenced housing reconstruction is community resilience, which can be explained as a neighborhood's network and action to promote neighborhood-based recovery. The author has interviewed neighborhood leaders in three neighborhoods, and found that neighborhood organizations play a significant role in neighborhood-based recovery, which might have some impact on individuals' housing reconstruction. We can well imagine that people cannot decide whether to come back to their neighborhood unless they know when the local school will start, when the levee restoration will be finished, if their neighbors are coming back to the devastated area, and so forth. It is not all about money, but the information of neighborhood recovery is needed for people's decision to come back and reconstruct their housing.

In Lakeview, residents who came back to the neighborhood provide disaster information to residents who evacuated out of town, and they continue to communicate with each other by utilizing mailing lists of the Lakeview Civic Improvement Association (LCIA). Through the media, LCIA sent messages to neighbors who were out of town that many of the neighbors were coming back to Lakeview, and this encouraged residents to come back because they understand the neighborhood's situation.

LCIA summarized, transmitted, and disseminated information regarding the disaster assistance program from government institutions and the Red Cross, as well as information of private contractors that have a direct relationship with the housing reconstruction. In addition to collecting and putting out information, LCIA advised neighbors to fill out an application form for the Road Home Program. LCIA worked as a "clearinghouse" for neighborhood recovery. Not only static socioeconomic factors such as vulnerability, but also dynamic socioeconomic factors such as community resilience might influence neighbors to return and reconstruct their housing.

The author is planning to conduct a questionnaire survey in early 2015, the 10th anniversary of Hurricane Katrina, for three neighborhoods' residents regarding their decision-making process of housing reconstruction; results will be published in a future research paper. "How did community resilience influence the individual?" is one question.

Checkerboard Housing Recovery

Through the annual survey, it was found that densely populated neighborhoods are coming back in bits and pieces, leaving some very sparsely populated areas, which can be called a checkerboard housing recovery [10]. What made this checkerboard situation? It is believed that this was triggered by a lack of a spatial planning perspective and neighborhood-based recovery strategy. First, the largest housing compensation program in U.S history, the Road Home Program, is designed around individual-based assistance and does not itself lead to neighborhood recovery. Second, there was no land use guidance or urban redevelopment projects put in place by the local government that had the perspective of a spatial planning regeneration approach at the neighborhood level. The only guidance for homeowners was to rebuild their homes along the elevation requirements of the National Flood Insurance Program as administered by FEMA.

Clustered Housing Recovery by Community Development Organization

There is one promising method to break through the checkerboard housing recovery situation and negative spiral: the Neighborhood Stabilization Program administered by NORA, working together with 14 consortium members, including housing developers and nonprofit housing developers, such as community development corporations. It is expected that communities that have suffered from foreclosures and abandonment are being stabilized through the purchase and redevelopment of foreclosed and abandoned housing and residential properties. This program holds the possibility to implement the Elevate and Cluster program proposed by the Unified New Orleans Plan (New Orleans recovery plan) to encourage residents to rebuild in clusters at higher elevations to help ensure vibrant neighborhoods and more efficient infrastructure investment in the context of a smaller overall population [11]. The expected impact of this program is not only improved safety, but also continuity of prior communities, housing rebuilding and restructuring of communities through flood-resistant designs, and the restoration of community services coordinated with individual housing reconstruction. The strategy used by Broadmoor Development Corporation, one of the consortium members, is unique: a combination of interventions and resources in selected zones targeted for development, including the renovation of occupied and vacant property, construction of infill development, and greening/maintenance of vacant lots [12].

The keywords of clustered housing recovery are *property transfer, property utilization*, and *community driven*. As mentioned in the introduction, clustered housing recovery is a housing reconstruction process controlled and guided by spatial planning, and it is considered effective in New Orleans in areas experiencing population decline before and after Hurricane Katrina, and reduces the risk of flood by relocating housing to safer areas. A positive aspect of this approach is that it can maximize the use of preexisting properties; for example, a property that was owned by homeowner A who opted not to rebuild will be redeveloped into new housing for homeowner B. In addition, homeowner A's property could be transformed into a commercial building or community facility, both of which are in high demand in the neighborhood. After a devastating disaster such as Hurricane Katrina, people cannot sustain their lives through housing recovery only; they also require other functions to support their livelihoods. That is, holistic disaster recovery is made possible by a process that is community driven. The vision of neighborhood recovery can be achieved through the use of property transfer, swapping, and clustering.

The concept of clustered housing recovery is considered effective in theory; however, there are many challenges ahead in its implementation. For example, preblighted properties that could be clustered are beyond NORA's control, and this made it difficult for NORA to utilize a sufficient number of properties to be clustered.

HOUSING RECOVERY AFTER GREAT EAST JAPAN EARTHQUAKE

The Great East Japan Earthquake of March 11, 2011, killed almost 18,000 people, and the following tsunami washed away more than 300 villages along the coast. Three years have passed since the disaster, now known as 3/11. One big difference between the United States and Japan is the leadership of national government in postdisaster recovery planning.

After 3/11, strict land use control for tsunami risk was enforced by the national government sector, and an unprecedented area is planned to be designated as hazardous, in which building is restricted [13]. One of the strong planning projects by the government was group relocation to mountainside areas in order to reduce tsunami risk. Survivors could participate in group relocation if their land was designated as hazardous. Also, the Japanese government is providing public housing for survivors who cannot afford to rebuild their housing, in contrast to the United States. There are multiple options for survivors in housing recovery: whether or not to reconstruct housing on their own, and whether or not to participate in group relocation.

Characteristics of Housing Recovery after 3/11

There were two major challenges for housing recovery after 3/11. One is that a significant percentage of housing would be controlled by postdisaster urban recovery projects, such as land readjustment projects with land raising and group relocation to mountainside areas. Strict land control for tsunami risk was enforced, and this is quite different from New Orleans post-Katrina, in which land use control has not improved despite experiencing a devastating natural disaster.

There is a commonality between the group residential relocation program after 3/11 and buyout programs after Hurricane Sandy. After Hurricane Sandy, which struck the East Coast of the United States in October 2012, New York City (NYC) decided to target flood risk zones and reduce rebuilding in these areas through the buyout of private properties. However, one major difference from the process in Japan is that the NYC government is not involved in the resettlement of residents. Residents themselves will decide where to move and resettle, which is the same as the phenomenon of individual relocation, described in the following section.

The other challenge is that the 3/11 tsunami experience forced survivors to decide whether to relocate or not. Even if local government didn't designate areas hazardous and forbidden for residential use, people are exploring where to live to avoid future tsunamis.

Individual Self-Help Housing Recovery with Relocation

The author particularly focuses attention on individual voluntary relocation and housing reconstruction that characterized the 3/11 housing recovery. Individual voluntary relocation is scattered spatially in the affected towns, which is similar to the phenomenon that occurred at the neighborhood scale in New Orleans after Katrina.

Based on housing rebuilding field surveys and mapping, individual voluntary relocation and housing reconstruction actions increased. Figure 12.10 shows the spatial distribution of newly constructed buildings in the city of Rikuzentakata, which indicates newly constructed housing after 3/11 [10]. Although all of these are not identified as the consequence of individual relocation, it is a high possibility, because it is unlikely that newcomers would move to the disaster-stricken area.

How did these individual self-help relocations result in the change of the urbanized area's spatial structure [14]? Looking in detail, these housing rebuilding patterns can be categorized in two ways: infill development type and scattered sprawl. Infill development housing recovery is considered to be an effective model to avoid low-density and scattering development following a devastating disaster. The scattered sprawl type leads to physically unsustainable recovery in a depopulated society, which triggers urban sprawl and low-density development. If there is a convenient residential area with a vacant lot, it can guide survivors to relocate.

Even if this housing rebuilding is physically unsustainable, these actions were the results of individuals deciding to sustain their living as quick as possible and avoid tsunami risk to achieve a feeling of safety

Figure 12.10 Spatial distribution of individual voluntary relocation after 3/11 in Rikuzentakata City.

[10]. Based on interviews with homeowners who decided to relocate individually, there were many survivors who decided against participating in group relocation and against reconstructing their housing on elevated land provided by the government. This is because they wanted to rebuild their housing as early as possible. Local government intended to resettle residents who lived in building restricted zones to new residential areas, but many residents dropped out of the process because of its slow speed.

Community-Driven Housing Recovery and Group Relocation after 3/11

The previous section illustrates that individual self-help relocation cannot create a physically sustainable neighborhood recovery without including a spatial planning perspective. Are there any alternatives to overcome this challenge? The author sees significant potential for community-driven group relocation in the Tohoku area after 3/11. There are two types of group relocation categorized by stakeholder involvement: government-driven group relocation and community-driven group relocation. In the former, local government takes leadership in finding land, negotiating with landowners for buyout, and coordinating the decision making among residents as a group. In the latter, residents form a group consisting of more than five households and assume the same roles as the government in the former type.

What is the difference between those two? Community-driven group relocation is not a reactive action based on choosing options provided by the government, but rather, it is a self-determining action among residents, not on an individual basis, about where to live and how to relocate. The practices of community-driven group recovery with infill development relocation in the city of Ofunato might have had a positive effect in geographical proximity, maintaining a sense of community, and sustaining people's living in order to achieve comprehensive sustainable disaster recovery. The only problem that individual relocation cannot resolve is how to sustain communities. This is why Japanese local governments want to conduct group relocation projects. If this decision-making process becomes community driven, it might be useful, as residents themselves can select where to live, direct communication with landowners can lead to a quicker overall process, and relocation sites would be closer to former homes, which enables them to sustain their community at a different neighborhood scale. What is important is not the type of development or group unit, but the community's self-determination; it is the most significant key to enabling these advantages for socially sustainable recovery.

Lack of Spatial Planning Perspective in Hurricane Katrina and Great East Japan Earthquake

What kind of housing recovery policy was prepared, and what were the outcomes and challenges ahead for housing recovery after the two catastrophes? It became clear that what both countries lacked in housing recovery was control and guidance for individual housing reconstruction (Figure 12.11). After Hurricane Katrina, local government did not assume leadership for land use planning, and the housing reconstruction subsidy program did not lead to neighborhood recovery. The checkerboard housing recovery situation at the neighborhood scale started a negative spiral that made it more difficult for residents to come back and newcomers to move into the area. In contrast to Hurricane Katrina, the recovery policy by the government sector in Japan after 3/11 took very strong leadership in setting hazardous zone building restrictions and promoting group relocation for residents who must relocate. However, these policies and programs did not work as the government expected. Homeowners are relocating to

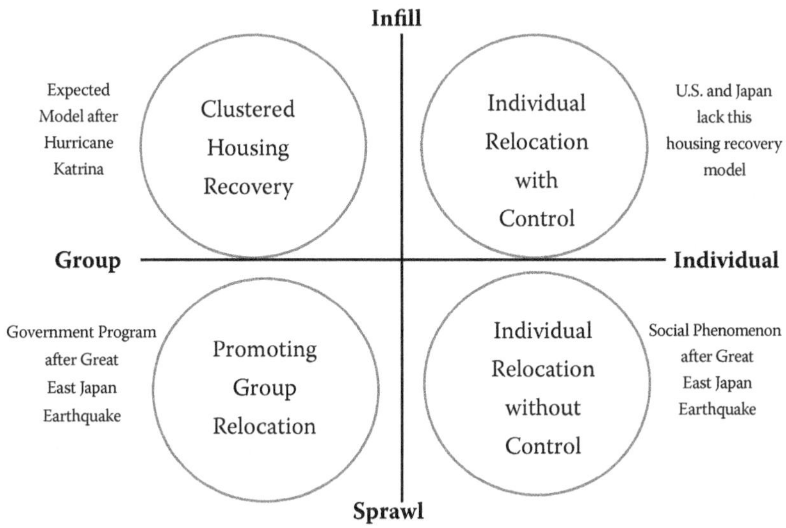

Figure 12.11 Housing recovery model by development type and stakeholder unit.

mountainside areas based on their individual decisions, not according to government plans. Individual self-help housing recovery with relocation will have a negative impact on physically sustainable disaster recovery at the citywide scale. Yet at the same time, this is a consequence of the resilience of homeowners who don't depend on government assistance.

CONCLUSION

This chapter gives an overview of postdisaster recovery phenomena and efforts after two natural catastrophes in terms of housing recovery and urban planning. It draws the conclusion that clustered housing recovery, from the perspective of spatial planning by utilizing survivors' resilience, can be suggested as an effective model for sustainable disaster recovery. Beyond using only urban development projects and control, it is important to apply urban planning guidance for housing recovery following a disaster. Specifically, local government can provide incentives for infill development relocation through housing subsidies.

Community-driven housing recovery has several positive aspects, such as the avoidance of a checkerboard recovery situation and scattered housing reconstruction, as well as many other advantages mentioned earlier in the text. The planning, program, policy, and process of housing recovery have to attain redundancy, which is one of the major components of the resiliency concept. Bruneau et al. (2003) state that resilience can be further defined as consisting of the following properties: robustness, redundancy, resourcefulness, and rapidity. Redundancy is defined as "the extent to which elements, systems, or other units of analysis exist that are substitutable, i.e., capable of satisfying functional requirements in the event of disruption, degradation, or loss of functionality" [15]. There are two redundancies that should be implemented in housing recovery. One is spatial redundancy, which can serve as vacant land to absorb and receive new housing reconstruction following a disaster. The second is stakeholder redundancy. After 3/11, we realized that too much dependence on seawalls to prevent tsunamis does not work, and the trend in Japanese society is toward approaches that embrace multiple measures for disaster reduction planning, like land use control and preparedness for evacuation. If we rely too much on one actor, such as the government sector, the speed of recovery will be delayed. Private housing developers are struggling to find land and develop residential areas to help survivors reconstruct housing. We have to utilize multiple stakeholders' roles, vitality, and energy in order to promote long-term disaster recovery. It is important to prepare multiple alternatives for housing recovery models, and establish the conditions in which residents can select the housing recovery model that they want the most.

ACKNOWLEDGMENTS

I express my gratitude to Dr. Elizabeth Maly (Tohoku University) and a number of Kobe University students who helped with the housing rebuilding survey and interviews in New Orleans. Housing rebuilding mapping after the Great East Japan Earthquake was made possible by the efforts of students at Kobe University and Meijo University. This research was supported by the Japan Society for the Promotion of Science, Grants-in-Aid for Scientific Research 25702021 (principal: Tamiyo Kondo) and 24310123 (principal: Yuka Karatani).

REFERENCES

1. Department of Housing and Urban Development (HUD). 2006. Current housing unit damage estimates: Hurricanes Katrina, Rita and Wilma.
2. National Climatic Data Center. 2005. Hurricane Katrina. December.
3. Vicki Mack and Allison Plyer. 2014. Neighborhood growth rates: Growth continues through 2014 in New Orleans neighborhoods. Greater New Orleans Data Center, August.
4. Robert B. Olshansky and Laurie Johnson. 2010. *Clear as Mud: Planning for the Rebuilding of New Orleans*. Washington, DC: American Planners Press, p. 170.
5. Frank Koughan. 2007. Housing and insurance. In *City Adrift: New Orleans Before and After Katrina*, ed. Jenni Bergal et al. Baton Rouge, LA: Louisiana State University Press, pp. 111–122.
6. Tamiyo Kondo and Titaya Sararit. 2012. Disaster mitigation and recovery planning after long-term flood: A case study of Hurricane Katrina (2005), 2011 Thailand flood, and 2011 Queensland flood. Presented at the Proceedings of the 9th Annual International Conference of the International Institute for Infrastructure Renewal and Reconstruction, Brisbane, Australia.
7. Timothy F. Green and Robert B. Olshansky. 2012. Rebuilding housing in New Orleans: The Road Home Program after the Hurricane Katrina disaster. *Housing Policy Debate*, 75–99.
8. Tamiyo Kondo. 2012. Housing rebuilding in the disaster area: A study on housing rebuilding in city of New Orleans after Hurricane Katrina (2005) [in Japanese]. *Journal of Architecture and Planning*, 679, 67–74.

9. Tamiyo Kondo. 2012. Annual change of housing rebuilding in the disaster area: A study on housing rebuilding in the city of New Orleans after Hurricane Katrina (2005) [in Japanese]. *Journal of Architecture and Planning,* 671, 2283–2292.
10. Tamiyo Kondo and Yuka Karatani. 2014. Housing recovery for sustainable disaster recovery through the case studies of Hurricane Katrina (2005) and the Great East Japan Earthquake (2011). In *Extended Abstract for 3rd International Conference on Urban Disaster Reduction,* October, Boulder, CO.
11. The Unified New Orleans Plan (UNOP): Citywide Strategic Recovery and Rebuilding Plan, 2007.
12. New Orleans Redevelopment Authority. 2013. Neighborhood Stabilization Program phase two report. July.
13. Kanako Iuchi. 2015. Disaster risk management and land use in Japan. In *Spatial Planning Following Disasters,* ed. S. Greiving et al. Bristol, UK: Policy Press.
14. Tamiyo Kondo and Yuka Karatani. 2015. Spatial planning control for housing recovery after the Great East Japan Earthquake. In *Spatial Planning Following Disasters,* ed. S. Greiving et al. Bristol, UK: Policy Press.
15. Michel Bruneau et al. 2003. A framework to quantitatively assess and enhance the seismic resilience of communities. *EERI Earthquake Spectra,* 19, 4.

INDEX

For Product Safety Concerns and Information please contact our EU representative GPSR@taylorandfrancis.com
Taylor & Francis Verlag GmbH, Kaufingerstraße 24, 80331 München, Germany

www.ingramcontent.com/pod-product-compliance
Lightning Source LLC
LaVergne TN
LVHW011113300926
849160LV00008B/370

* 9 7 8 1 4 8 2 2 4 7 4 0 4 *